Thermodynamic and Transport Properties of Fluids

SI UNITS

FIFTH EDITION

ARRANGED BY

G. F. C. Rogers and Y. R. Mayhew

Blackwell
Publishing

Editorial offices:
Blackwell Publishing Ltd, 9600 Garsington Road, Oxford OX4 2DQ, UK
Tel: +44 (0) 1865 776868
Blackwell Publishing Inc., 350 Main Street, Malden, MA 02148-5020, USA
Tel: +1 781 388 8250
Blackwell Publishing Asia Pty Ltd, 550 Swanston Street, Carlton, Victoria 3053, Australia
Tel: +61 (0)3 8359 1011

First published 1964 by Blackwell Publishing Ltd
Second edition 1967
Third edition 1980
Fourth edition 1988
Fifth edition 1995

28   2015

ISBN: 978-0-631-19703-4

Library of Congress Cataloging-in-Publication Data is available

A catalogue record for this title is available from the British Library

Set by Advanced Filmsetters (Glasgow) Ltd
Updates set by Graphicraft Limited, Hong Kong
Printed and bound in Great Britain
by TJ International Ltd, Padstow, Cornwall

The publisher's policy is to use permanent paper from mills that operate a sustainable
forestry policy, and which has been manufactured from pulp processed using acid-free and
elementary chlorine-free practices. Furthermore, the publisher ensures that the text paper and
cover board used have met acceptable environmental accreditation standards.

For further information on
Blackwell Publishing, visit our website:
www.blackwellpublishing.com

CONTENTS

NOTATION AND UNITS

a	m/s	– velocity of sound
$c_p, \tilde{c}_p$	kJ/kg K, kJ/kmol K	– specific, molar heat capacity at constant p
$c_v, \tilde{c}_v$	kJ/kg K, kJ/kmol K	– specific, molar heat capacity at constant v
$g, \tilde{g}$	kJ/kg, kJ/kmol	– specific, molar Gibbs function $(h - Ts, \tilde{h} - T\tilde{s})$
$\Delta \tilde{g}^{\ominus}, \Delta g_f^{\ominus}$	kJ/kmol	– molar Gibbs function of reaction, of formation
$h, \tilde{h}$	kJ/kg, kJ/kmol	– specific, molar enthalpy $(u + pv, \tilde{u} + p\tilde{v})$
$\Delta \tilde{h}^{\ominus}, \Delta h_f^{\ominus}$	kJ/kmol	– molar enthalpy of reaction, of formation
$K^{\ominus}, K_f^{\ominus}$	–	– equilibrium constant, of formation
k	kW/m K	– thermal conductivity
$\tilde{m}$	kg/kmol	– molar mass
p	bar	– absolute pressure
Pr	–	– Prandtl number $(c_p \mu / k)$
$R, \tilde{R}$	kJ/kg K, kJ/kmol K	– specific, molar (universal) gas constant
$s, \tilde{s}$	kJ/kg K, kJ/kmol K	– specific, molar entropy
T	K or °C	–absolute temperature (K) or Celsius temperature (°C)
ΔT	K	–temperature interval or difference
$u, \tilde{u}$	kJ/kg, kJ/kmol	– specific, molar internal energy
$v, \tilde{v}$	m³/kg, m³/kmol	– specific, molar volume $(1/\rho, 1/\tilde{\rho})$
z	m	– geometric altitude above sea level
γ	–	– ratio of specific heat capacities $(c_p/c_v = \tilde{c}_p/\tilde{c}_v)$
λ	m	– mean free path
μ	kg/m s = N s/m²	– dynamic viscosity
ν	m²/s	– kinematic viscosity (μ/ρ)
$\rho, \tilde{\rho}$	kg/m³, kmol/m³	– mass, molar density $(1/v, 1/\tilde{v})$

Subscripts

c   – refers to a property in the critical state
f   – refers to a property of the saturated liquid, or to a value of formation
g   – refers to a property of the saturated vapour
fg – refers to a change of phase at constant p
i   – refers to a property of the saturated solid
s   – refers to a saturation temperature or pressure

Superscripts

~   – refers to a molar property (i.e. per unit amount-of-substance)
⊖   – refers to a property at standard pressure $p^{\ominus} = 1$ bar (the superscript o is often used)

Saturated Water and Steam

T [°C]	p_s [bar]	v_g [m³/kg]	h_f	h_{fg} [kJ/kg]	h_g	s_f	s_{fg} [kJ/kg K]	s_g
0.01	0.006112	206.1	0*	2500.8	2500.8	0†	9.155	9.155
1	0.006566	192.6	4.2	2498.3	2502.5	0.015	9.113	9.128
2	0.007054	179.9	8.4	2495.9	2504.3	0.031	9.071	9.102
3	0.007575	168.2	12.6	2493.6	2506.2	0.046	9.030	9.076
4	0.008129	157.3	16.8	2491.3	2508.1	0.061	8.989	9.050
5	0.008719	147.1	21.0	2488.9	2509.9	0.076	8.948	9.024
6	0.009346	137.8	25.2	2486.6	2511.8	0.091	8.908	8.999
7	0.01001	129.1	29.4	2484.3	2513.7	0.106	8.868	8.974
8	0.01072	121.0	33.6	2481.9	2515.5	0.121	8.828	8.949
9	0.01147	113.4	37.8	2479.6	2517.4	0.136	8.788	8.924
10	0.01227	106.4	42.0	2477.2	2519.2	0.151	8.749	8.900
11	0.01312	99.90	46.2	2474.9	2521.1	0.166	8.710	8.876
12	0.01401	93.83	50.4	2472.5	2522.9	0.180	8.671	8.851
13	0.01497	88.17	54.6	2470.2	2524.8	0.195	8.633	8.828
14	0.01597	82.89	58.8	2467.8	2526.6	0.210	8.594	8.804
15	0.01704	77.97	62.9	2465.5	2528.4	0.224	8.556	8.780
16	0.01817	73.38	67.1	2463.1	2530.2	0.239	8.518	8.757
17	0.01936	69.09	71.3	2460.8	2532.1	0.253	8.481	8.734
18	0.02063	65.08	75.5	2458.4	2533.9	0.268	8.444	8.712
19	0.02196	61.34	79.7	2456.0	2535.7	0.282	8.407	8.689
20	0.02337	57.84	83.9	2453.7	2537.6	0.296	8.370	8.666
21	0.02486	54.56	88.0	2451.4	2539.4	0.310	8.334	8.644
22	0.02642	51.49	92.2	2449.0	2541.2	0.325	8.297	8.622
23	0.02808	48.62	96.4	2446.6	2543.0	0.339	8.261	8.600
24	0.02982	45.92	100.6	2444.2	2544.8	0.353	8.226	8.579
25	0.03166	43.40	104.8	2441.8	2546.6	0.367	8.190	8.557
26	0.03360	41.03	108.9	2439.5	2548.4	0.381	8.155	8.536
27	0.03564	38.81	113.1	2437.2	2550.3	0.395	8.120	8.515
28	0.03778	36.73	117.3	2434.8	2552.1	0.409	8.085	8.494
29	0.04004	34.77	121.5	2432.4	2553.9	0.423	8.050	8.473
30	0.04242	32.93	125.7	2430.0	2555.7	0.436	8.016	8.452
32	0.04754	29.57	134.0	2425.3	2559.3	0.464	7.948	8.412
34	0.05318	26.60	142.4	2420.5	2562.9	0.491	7.881	8.372
36	0.05940	23.97	150.7	2415.8	2566.5	0.518	7.814	8.332
38	0.06624	21.63	159.1	2411.0	2570.1	0.545	7.749	8.294
40	0.07375	19.55	167.5	2406.2	2573.7	0.572	7.684	8.256
42	0.08198	17.69	175.8	2401.4	2577.2	0.599	7.620	8.219
44	0.09100	16.03	184.2	2396.6	2580.8	0.625	7.557	8.182
46	0.1009	14.56	192.5	2391.8	2584.3	0.651	7.494	8.145
48	0.1116	13.23	200.9	2387.0	2587.9	0.678	7.433	8.111
50	0.1233	12.04	209.3	2382.1	2591.4	0.704	7.371	8.075
55	0.1574	9.578	230.2	2370.1	2600.3	0.768	7.223	7.991
60	0.1992	7.678	251.1	2357.9	2609.0	0.831	7.078	7.909
65	0.2501	6.201	272.0	2345.7	2617.7	0.893	6.937	7.830
70	0.3116	5.045	293.0	2333.3	2626.3	0.955	6.800	7.755
75	0.3855	4.133	313.9	2320.8	2634.7	1.015	6.666	7.681
80	0.4736	3.408	334.9	2308.3	2643.2	1.075	6.536	7.611
85	0.5780	2.828	355.9	2295.6	2651.5	1.134	6.410	7.544
90	0.7011	2.361	376.9	2282.8	2659.7	1.192	6.286	7.478
95	0.8453	1.982	398.0	2269.8	2667.8	1.250	6.166	7.416
100	1.01325	1.673	419.1	2256.7	2675.8	1.307	6.048	7.355

† u and s are chosen to be zero for saturated liquid at the triple point.

Note: values of v_f can be found on p. 10.

p	T_s	v_g	u_f	u_g	h_f	h_{fg}	h_g	s_f	s_{fg}	s_g
[bar]	[°C]	[m³/kg]	[kJ/kg]		[kJ/kg]			[kJ/kg K]		
0.006112	0.01	206.1	0†	2375	0*	2501	2501	0†	9.155	9.155
0.010	7.0	129.2	29	2385	29	2485	2514	0.106	8.868	8.974
0.015	13.0	87.98	55	2393	55	2470	2525	0.196	8.631	8.827
0.020	17.5	67.01	73	2399	73	2460	2533	0.261	8.462	8.723
0.025	21.1	54.26	88	2403	88	2451	2539	0.312	8.330	8.642
0.030	24.1	45.67	101	2408	101	2444	2545	0.354	8.222	8.576
0.035	26.7	39.48	112	2412	112	2438	2550	0.391	8.130	8.521
0.040	29.0	34.80	121	2415	121	2433	2554	0.422	8.051	8.473
0.045	31.0	31.14	130	2418	130	2428	2558	0.451	7.980	8.431
0.050	32.9	28.20	138	2420	138	2423	2561	0.476	7.918	8.394
0.055	34.6	25.77	145	2422	145	2419	2564	0.500	7.860	8.360
0.060	36.2	23.74	152	2425	152	2415	2567	0.521	7.808	8.329
0.065	37.7	22.02	158	2427	158	2412	2570	0.541	7.760	8.301
0.070	39.0	20.53	163	2428	163	2409	2572	0.559	7.715	8.274
0.075	40.3	19.24	169	2430	169	2405	2574	0.576	7.674	8.250
0.080	41.5	18.10	174	2432	174	2402	2576	0.593	7.634	8.227
0.085	42.7	17.10	179	2434	179	2400	2579	0.608	7.598	8.206
0.090	43.8	16.20	183	2435	183	2397	2580	0.622	7.564	8.186
0.095	44.8	15.40	188	2436	188	2394	2582	0.636	7.531	8.167
0.100	45.8	14.67	192	2437	192	2392	2584	0.649	7.500	8.149
0.12	49.4	12.36	207	2442	207	2383	2590	0.696	7.389	8.085
0.14	52.6	10.69	220	2446	220	2376	2596	0.737	7.294	8.031
0.16	55.3	9.432	232	2450	232	2369	2601	0.772	7.213	7.985
0.18	57.8	8.444	242	2453	242	2363	2605	0.804	7.140	7.944
0.20	60.1	7.648	251	2456	251	2358	2609	0.832	7.075	7.907
0.22	62.2	6.994	260	2459	260	2353	2613	0.858	7.016	7.874
0.24	64.1	6.445	268	2461	268	2348	2616	0.882	6.962	7.844
0.26	65.9	5.979	276	2464	276	2343	2619	0.904	6.913	7.817
0.28	67.5	5.578	283	2466	283	2339	2622	0.925	6.866	7.791
0.30	69.1	5.228	289	2468	289	2336	2625	0.944	6.823	7.767
0.32	70.6	4.921	295	2470	295	2332	2627	0.962	6.783	7.745
0.34	72.0	4.649	302	2472	302	2328	2630	0.980	6.745	7.725
0.36	73.4	4.407	307	2473	307	2325	2632	0.996	6.709	7.705
0.38	74.7	4.189	312	2475	312	2322	2634	1.011	6.675	7.686
0.40	75.9	3.992	318	2476	318	2318	2636	1.026	6.643	7.669
0.42	77.1	3.814	323	2478	323	2315	2638	1.040	6.612	7.652
0.44	78.2	3.651	327	2479	327	2313	2640	1.054	6.582	7.636
0.46	79.3	3.502	332	2481	332	2310	2642	1.067	6.554	7.621
0.48	80.3	3.366	336	2482	336	2308	2644	1.079	6.528	7.607
0.50	81.3	3.239	340	2483	340	2305	2645	1.091	6.502	7.593
0.55	83.7	2.964	351	2486	351	2298	2649	1.119	6.442	7.561
0.60	86.0	2.731	360	2489	360	2293	2653	1.145	6.386	7.531
0.65	88.0	2.535	369	2492	369	2288	2657	1.169	6.335	7.504
0.70	90.0	2.364	377	2494	377	2283	2660	1.192	6.286	7.478
0.75	91.8	2.217	384	2496	384	2278	2662	1.213	6.243	7.456
0.80	93.5	2.087	392	2498	392	2273	2665	1.233	6.201	7.434
0.85	95.2	1.972	399	2500	399	2269	2668	1.252	6.162	7.414
0.90	96.7	1.869	405	2502	405	2266	2671	1.270	6.124	7.394
0.95	98.2	1.777	411	2504	411	2262	2673	1.287	6.089	7.376
1.00	99.6	1.694	417	2506	417	2258	2675	1.303	6.056	7.359

$$* \quad \frac{h_f}{[kJ/kg]} = \frac{pv_f}{[kJ/kg]} = \frac{p}{[bar]} \times \frac{10^5[N]}{[m^2]} \times \frac{v_f}{[m^3/kg]} \times \left[\frac{m^3}{kg}\right] \times \frac{[kJ]}{10^3[Nm]} \times \frac{1}{[kJ/kg]}$$

$$= \frac{p}{[bar]} \times \frac{v_f}{[m^3/kg]} \times 10^2 = 0.006112 \times 0.0010002 \times 10^2 = 0.0006112$$

3

Saturated Water and Steam

p [bar]	T_s [°C]	v_g [m^3/kg]	u_f [kJ/kg]	u_g [kJ/kg]	h_f [kJ/kg]	h_{fg} [kJ/kg]	h_g [kJ/kg]	s_f [kJ/kg K]	s_{fg} [kJ/kg K]	s_g [kJ/kg K]
1.0	99.6	1.694	417	2506	417	2258	2675	1.303	6.056	7.359
1.1	102.3	1.549	429	2510	429	2251	2680	1.333	5.994	7.327
1.2	104.8	1.428	439	2512	439	2244	2683	1.361	5.937	7.298
1.3	107.1	1.325	449	2515	449	2238	2687	1.387	5.884	7.271
1.4	109.3	1.236	458	2517	458	2232	2690	1.411	5.835	7.246
1.5	111.4	1.159	467	2519	467	2226	2693	1.434	5.789	7.223
1.6	113.3	1.091	475	2521	475	2221	2696	1.455	5.747	7.202
1.7	115.2	1.031	483	2524	483	2216	2699	1.475	5.707	7.182
1.8	116.9	0.9774	491	2526	491	2211	2702	1.494	5.669	7.163
1.9	118.6	0.9292	498	2528	498	2206	2704	1.513	5.632	7.145
2.0	120.2	0.8856	505	2530	505	2202	2707	1.530	5.597	7.127
2.1	121.8	0.8461	511	2531	511	2198	2709	1.547	5.564	7.111
2.2	123.3	0.8100	518	2533	518	2193	2711	1.563	5.533	7.096
2.3	124.7	0.7770	524	2534	524	2189	2713	1.578	5.503	7.081
2.4	126.1	0.7466	530	2536	530	2185	2715	1.593	5.474	7.067
2.5	127.4	0.7186	535	2537	535	2182	2717	1.607	5.446	7.053
2.6	128.7	0.6927	541	2539	541	2178	2719	1.621	5.419	7.040
2.7	130.0	0.6686	546	2540	546	2174	2720	1.634	5.393	7.027
2.8	131.2	0.6462	551	2541	551	2171	2722	1.647	5.368	7.015
2.9	132.4	0.6253	556	2543	556	2168	2724	1.660	5.344	7.004
3.0	133.5	0.6057	561	2544	561	2164	2725	1.672	5.321	6.993
3.5	138.9	0.5241	584	2549	584	2148	2732	1.727	5.214	6.941
4.0	143.6	0.4623	605	2554	605	2134	2739	1.776	5.121	6.897
4.5	147.9	0.4139	623	2558	623	2121	2744	1.820	5.037	6.857
5.0	151.8	0.3748	639	2562	640	2109	2749	1.860	4.962	6.822
5.5	155.5	0.3427	655	2565	656	2097	2753	1.897	4.893	6.790
6	158.8	0.3156	669	2568	670	2087	2757	1.931	4.830	6.761
7	165.0	0.2728	696	2573	697	2067	2764	1.992	4.717	6.709
8	170.4	0.2403	720	2577	721	2048	2769	2.046	4.617	6.663
9	175.4	0.2149	742	2581	743	2031	2774	2.094	4.529	6.623
10	179.9	0.1944	762	2584	763	2015	2778	2.138	4.448	6.586
11	184.1	0.1774	780	2586	781	2000	2781	2.179	4.375	6.554
12	188.0	0.1632	797	2588	798	1986	2784	2.216	4.307	6.523
13	191.6	0.1512	813	2590	815	1972	2787	2.251	4.244	6.495
14	195.0	0.1408	828	2593	830	1960	2790	2.284	4.185	6.469
15	198.3	0.1317	843	2595	845	1947	2792	2.315	4.130	6.445
16	201.4	0.1237	857	2596	859	1935	2794	2.344	4.078	6.422
17	204.3	0.1167	870	2597	872	1923	2795	2.372	4.028	6.400
18	207.1	0.1104	883	2598	885	1912	2797	2.398	3.981	6.379
19	209.8	0.1047	895	2599	897	1901	2798	2.423	3.936	6.359
20	212.4	0.09957	907	2600	909	1890	2799	2.447	3.893	6.340
22	217.2	0.09069	928	2601	931	1870	2801	2.492	3.813	6.305
24	221.8	0.08323	949	2602	952	1850	2802	2.534	3.738	6.272
26	226.0	0.07689	969	2603	972	1831	2803	2.574	3.668	6.242
28	230.0	0.07142	988	2603	991	1812	2803	2.611	3.602	6.213
30	233.8	0.06665	1004	2603	1008	1795	2803	2.645	3.541	6.186
32	237.4	0.06246	1021	2603	1025	1778	2803	2.679	3.482	6.161
34	240.9	0.05875	1038	2603	1042	1761	2803	2.710	3.426	6.136
36	244.2	0.05544	1054	2602	1058	1744	2802	2.740	3.373	6.113
38	247.3	0.05246	1068	2602	1073	1729	2802	2.769	3.322	6.091
40	250.3	0.04977	1082	2602	1087	1714	2801	2.797	3.273	6.070

4

p	T_s	v_g	u_f	u_g	h_f	h_{fg}	h_g	s_f	s_{fg}	s_g
[bar]	[°C]	[m³/kg]	[kJ/kg]		[kJ/kg]			[kJ/kg K]		
40	250.3	0.04977	1082	2602	1087	1714	2801	2.797	3.273	6.070
42	253.2	0.04732	1097	2601	1102	1698	2800	2.823	3.226	6.049
44	256.0	0.04509	1109	2600	1115	1683	2798	2.849	3.180	6.029
46	258.8	0.04305	1123	2599	1129	1668	2797	2.874	3.136	6.010
48	261.4	0.04117	1136	2598	1142	1654	2796	2.897	3.094	5.991
50	263.9	0.03944	1149	2597	1155	1639	2794	2.921	3.052	5.973
55	269.9	0.03563	1178	2594	1185	1605	2790	2.976	2.955	5.931
60	275.6	0.03244	1206	2590	1214	1570	2784	3.027	2.863	5.890
65	280.8	0.02972	1232	2586	1241	1538	2779	3.076	2.775	5.851
70	285.8	0.02737	1258	2581	1267	1505	2772	3.122	2.692	5.814
75	290.5	0.02532	1283	2576	1293	1473	2766	3.166	2.613	5.779
80	295.0	0.02352	1306	2570	1317	1441	2758	3.207	2.537	5.744
85	299.2	0.02192	1329	2565	1341	1410	2751	3.248	2.463	5.711
90	303.3	0.02048	1351	2559	1364	1379	2743	3.286	2.393	5.679
95	307.2	0.01919	1372	2552	1386	1348	2734	3.324	2.323	5.647
100	311.0	0.01802	1393	2545	1408	1317	2725	3.360	2.255	5.615
105	314.6	0.01696	1414	2537	1429	1286	2715	3.395	2.189	5.584
110	318.0	0.01598	1434	2529	1450	1255	2705	3.430	2.123	5.553
115	321.4	0.01508	1454	2522	1471	1224	2695	3.463	2.060	5.523
120	324.6	0.01426	1473	2514	1491	1194	2685	3.496	1.997	5.493
125	327.8	0.01349	1492	2505	1511	1163	2674	3.529	1.934	5.463
130	330.8	0.01278	1511	2496	1531	1131	2662	3.561	1.872	5.433
135	333.8	0.01211	1530	2487	1551	1099	2650	3.592	1.811	5.403
140	336.6	0.01149	1548	2477	1571	1067	2638	3.623	1.750	5.373
145	339.4	0.01090	1567	2467	1591	1034	2625	3.654	1.689	5.343
150	342.1	0.01035	1585	2456	1610	1001	2611	3.685	1.627	5.312
155	344.8	0.00982	1604	2445	1630	967	2597	3.715	1.565	5.280
160	347.3	0.00932	1623	2433	1650	932	2582	3.746	1.502	5.248
165	349.8	0.00884	1641	2420	1670	895	2565	3.777	1.437	5.214
170	352.3	0.00838	1660	2406	1690	858	2548	3.808	1.373	5.181
175	354.6	0.00794	1679	2391	1711	819	2530	3.839	1.305	5.144
180	357.0	0.00751	1699	2375	1732	778	2510	3.872	1.236	5.108
185	359.2	0.00709	1719	2358	1754	735	2489	3.905	1.163	5.068
190	361.4	0.00668	1740	2339	1777	689	2466	3.941	1.086	5.027
195	363.6	0.00627	1762	2318	1801	639	2440	3.977	1.004	4.981
200	365.7	0.00585	1786	2294	1827	584	2411	4.014	0.914	4.928
202	366.5	0.00569	1796	2283	1838	560	2398	4.031	0.875	4.906
204	367.4	0.00552	1806	2271	1849	535	2384	4.049	0.835	4.884
206	368.2	0.00534	1817	2259	1861	508	2369	4.067	0.792	4.859
208	369.0	0.00517	1829	2245	1874	479	2353	4.087	0.745	4.832
210	369.8	0.00498	1842	2231	1889	447	2336	4.108	0.695	4.803
212	370.6	0.00479	1856	2214	1904	412	2316	4.131	0.640	4.771
214	371.4	0.00458	1871	2196	1921	373	2294	4.157	0.579	4.736
216	372.1	0.00436	1888	2174	1940	328	2268	4.186	0.508	4.694
218	372.9	0.00409	1911	2146	1965	270	2235	4.224	0.417	4.641
220	373.7	0.00368	1949	2097	2008	170	2178	4.289	0.263	4.552
221.2	374.15	0.00317	2014	2014	2084	0	2084	4.406	0.000	4.406

Superheated Steam†

p/[bar] (Ts/[°C])		T/[°C]	50	100	150	200	250	300	400	500
0	$u = h - RT$ at $p = 0$	v								
		u	2446	2517	2589	2662	2737	2812	2969	3132
		h	2595	2689	2784	2880	2978	3077	3280	3489
		s								
0.006112 (0.01)	v_g 206.1	v	243.9	281.7	319.5	357.3	395.0	432.8	508.3	583.8
	u_g 2375	u	2446	2517	2589	2662	2737	2812	2969	3132
	h_g 2501	h	2595	2689	2784	2880	2978	3077	3280	3489
	s_g 9.155	s	9.468	9.739	9.978	10.193	10.390	10.571	10.897	11.187
0.01 (7.0)	v_g 129.2	v	149.1	172.2	195.3	218.4	241.4	264.5	310.7	356.8
	u_g 2385	u	2446	2517	2589	2662	2737	2812	2969	3132
	h_g 2514	h	2595	2689	2784	2880	2978	3077	3280	3489
	s_g 8.974	s	9.241	9.512	9.751	9.966	10.163	10.344	10.670	10.960
0.05 (32.9)	v_g 28.20	v	29.78	34.42	39.04	43.66	48.28	52.90	62.13	71.36
	u_g 2420	u	2445	2516	2589	2662	2737	2812	2969	3132
	h_g 2561	h	2594	2688	2784	2880	2978	3077	3280	3489
	s_g 8.394	s	8.496	8.768	9.008	9.223	9.420	9.601	9.927	10.217
0.1 (45.8)	v_g 14.67	v	14.87	17.20	19.51	21.83	24.14	26.45	31.06	35.68
	u_g 2437	u	2443	2516	2588	2662	2736	2812	2969	3132
	h_g 2584	h	2592	2688	2783	2880	2977	3077	3280	3489
	s_g 8.149	s	8.173	8.447	8.688	8.903	9.100	9.281	9.607	9.897
0.5 (81.3)	v_g 3.239	v		3.420	3.890	4.356	4.821	5.284	6.209	7.134
	u_g 2483	u		2512	2585	2660	2735	2812	2969	3132
	h_g 2645	h		2683	2780	2878	2976	3076	3279	3489
	s_g 7.593	s		7.694	7.940	8.158	8.355	8.537	8.864	9.154
0.75 (91.8)	v_g 2.217	v		2.271	2.588	2.901	3.211	3.521	4.138	4.755
	u_g 2496	u		2510	2585	2659	2734	2811	2969	3132
	h_g 2662	h		2680	2779	2877	2975	3075	3279	3489
	s_g 7.456	s		7.500	7.750	7.969	8.167	8.349	8.676	8.967
1 (99.6)	v_g 1.694	v		1.696	1.937	2.173	2.406	2.639	3.103	3.565
	u_g 2506	u		2506	2583	2659	2734	2811	2968	3131
	h_g 2675	h		2676	2777	2876	2975	3075	3278	3488
	s_g 7.359	s		7.360	7.614	7.834	8.033	8.215	8.543	8.834
1.01325 (100.0)	v_g 1.673	v			1.912	2.145	2.375	2.604	3.062	3.519
	u_g 2506	u			2583	2659	2734	2811	2968	3131
	h_g 2676	h			2777	2876	2975	3075	3278	3488
	s_g 7.355	s			7.608	7.828	8.027	8.209	8.537	8.828
1.5 (111.4)	v_g 1.159	v			1.286	1.445	1.601	1.757	2.067	2.376
	u_g 2519	u			2580	2656	2733	2809	2967	3131
	h_g 2693	h			2773	2873	2973	3073	3277	3488
	s_g 7.223	s			7.420	7.643	7.843	8.027	8.355	8.646
2 (120.2)	v_g 0.8856	v			0.9602	1.081	1.199	1.316	1.549	1.781
	u_g 2530	u			2578	2655	2731	2809	2967	3131
	h_g 2707	h			2770	2871	2971	3072	3277	3487
	s_g 7.127	s			7.280	7.507	7.708	7.892	8.221	8.513
3 (133.5)	v_g 0.6057	v			0.6342	0.7166	0.7965	0.8754	1.031	1.187
	u_g 2544	u			2572	2651	2729	2807	2966	3130
	h_g 2725	h			2762	2866	2968	3070	3275	3486
	s_g 6.993	s			7.078	7.312	7.517	7.702	8.032	8.324
4 (143.6)	v_g 0.4623	$v/[\mathrm{m^3/kg}]$			0.4710	0.5345	0.5953	0.6549	0.7725	0.8893
	u_g 2554	$u/[\mathrm{kJ/kg}]$			2565	2648	2727	2805	2965	3129
	h_g 2739	$h/[\mathrm{kJ/kg}]$			2753	2862	2965	3067	3274	3485
	s_g 6.897	$s/[\mathrm{kJ/kg\,K}]$			6.929	7.172	7.379	7.566	7.898	8.191

† The entries in all tables are regarded as pure numbers and therefore the symbols for the physical quantities should be divided by the appropriate units as shown for the entries at $p/[\mathrm{bar}] = 4$. Because of lack of space, this has not been done consistently in the superheat and supercritical tables on pp. 6–9 and in the tables on pp. 11 and 23.

$p/[\text{bar}]$ $(T_s/[°C])$			$\dfrac{T}{[°C]}$	200	250	300	350	400	450	500	600
5 (151.8)	v_g	0.3748	v	0.4252	0.4745	0.5226	0.5701	0.6172	0.6641	0.7108	0.8040
	u_g	2562	u	2644	2725	2804	2883	2963	3045	3129	3300
	h_g	2749	h	2857	2962	3065	3168	3272	3377	3484	3702
	s_g	6.822	s	7.060	7.271	7.460	7.633	7.793	7.944	8.087	8.351
6 (158.8)	v_g	0.3156	v	0.3522	0.3940	0.4344	0.4743	0.5136	0.5528	0.5919	0.6697
	u_g	2568	u	2640	2722	2801	2881	2962	3044	3128	3299
	h_g	2757	h	2851	2958	3062	3166	3270	3376	3483	3701
	s_g	6.761	s	6.968	7.182	7.373	7.546	7.707	7.858	8.001	8.267
7 (165.0)	v_g	0.2728	v	0.3001	0.3364	0.3714	0.4058	0.4397	0.4734	0.5069	0.5737
	u_g	2573	u	2636	2720	2800	2880	2961	3043	3127	3298
	h_g	2764	h	2846	2955	3060	3164	3269	3374	3482	3700
	s_g	6.709	s	6.888	7.106	7.298	7.473	7.634	7.786	7.929	8.195
8 (170.4)	v_g	0.2403	v	0.2610	0.2933	0.3242	0.3544	0.3842	0.4138	0.4432	0.5018
	u_g	2577	u	2631	2716	2798	2878	2960	3042	3126	3298
	h_g	2769	h	2840	2951	3057	3162	3267	3373	3481	3699
	s_g	6.663	s	6.817	7.040	7.233	7.409	7.571	7.723	7.866	8.132
9 (175.4)	v_g	0.2149	v	0.2305	0.2597	0.2874	0.3144	0.3410	0.3674	0.3937	0.4458
	u_g	2581	u	2628	2714	2796	2877	2959	3041	3126	3298
	h_g	2774	h	2835	2948	3055	3160	3266	3372	3480	3699
	s_g	6.623	s	6.753	6.980	7.176	7.352	7.515	7.667	7.811	8.077
10 (179.9)	v_g	0.1944	v	0.2061	0.2328	0.2580	0.2825	0.3065	0.3303	0.3540	0.4010
	u_g	2584	u	2623	2711	2794	2875	2957	3040	3124	3297
	h_g	2778	h	2829	2944	3052	3158	3264	3370	3478	3698
	s_g	6.586	s	6.695	6.926	7.124	7.301	7.464	7.617	7.761	8.028
15 (198.3)	v_g	0.1317	v	0.1324	0.1520	0.1697	0.1865	0.2029	0.2191	0.2351	0.2667
	u_g	2595	u	2597	2697	2784	2868	2952	3035	3120	3294
	h_g	2792	h	2796	2925	3039	3148	3256	3364	3473	3694
	s_g	6.445	s	6.452	6.711	6.919	7.102	7.268	7.423	7.569	7.838
20 (212.4)	v_g	0.0996	v		0.1115	0.1255	0.1386	0.1511	0.1634	0.1756	0.1995
	u_g	2600	u		2681	2774	2861	2946	3030	3116	3291
	h_g	2799	h		2904	3025	3138	3248	3357	3467	3690
	s_g	6.340	s		6.547	6.768	6.957	7.126	7.283	7.431	7.701
30 (233.8)	v_g	0.0666	v		0.0706	0.0812	0.0905	0.0993	0.1078	0.1161	0.1324
	u_g	2603	u		2646	2751	2845	2933	3020	3108	3285
	h_g	2803	h		2858	2995	3117	3231	3343	3456	3682
	s_g	6.186	s		6.289	6.541	6.744	6.921	7.082	7.233	7.507
40 (250.3)	v_g	0.0498	v			0.0588	0.0664	0.0733	0.0800	0.0864	0.0988
	u_g	2602	u			2728	2828	2921	3010	3099	3279
	h_g	2801	h			2963	3094	3214	3330	3445	3674
	s_g	6.070	s			6.364	6.584	6.769	6.935	7.089	7.368
50 (263.9)	v_g	0.0394	v			0.0453	0.0519	0.0578	0.0632	0.0685	0.0786
	u_g	2597	u			2700	2810	2907	3000	3090	3273
	h_g	2794	h			2927	3070	3196	3316	3433	3666
	s_g	5.973	s			6.212	6.451	6.646	6.818	6.975	7.258
60 (275.6)	v_g	0.0324	v			0.0362	0.0422	0.0473	0.0521	0.0566	0.0652
	u_g	2590	u			2670	2792	2893	2988	3081	3266
	h_g	2784	h			2887	3045	3177	3301	3421	3657
	s_g	5.890	s			6.071	6.336	6.541	6.719	6.879	7.166
70 (285.8)	v_g	0.0274	$v/[\text{m}^3/\text{kg}]$			0.0295	0.0352	0.0399	0.0441	0.0481	0.0556
	u_g	2581	$u/[\text{kJ}/\text{kg}]$			2634	2772	2879	2978	3073	3260
	h_g	2772	$h/[\text{kJ}/\text{kg}]$			2841	3018	3158	3287	3410	3649
	s_g	5.814	$s/[\text{kJ}/\text{kg K}]$			5.934	6.231	6.448	6.632	6.796	7.088

* See footnote on p. 6.

Superheated Steam*

$p/[\text{bar}]$ ($T_r/[°C]$)		$\dfrac{T}{[°C]}$	350	375	400	425	450	500	600	700
80 (295.0)	v_g 0.02352	$v/10^{-2}$	2.994	3.220	3.428	3.625	3.812	4.170	4.839	5.476
	h_g 2758	h	2990	3067	3139	3207	3272	3398	3641	3881
	s_g 5.744	s	6.133	6.255	6.364	6.463	6.555	6.723	7.019	7.279
90 (303.3)	v_g 0.02048	$v/10^{-2}$	2.578	2.794	2.991	3.173	3.346	3.673	4.279	4.852
	h_g 2743	h	2959	3042	3118	3189	3256	3385	3633	3874
	s_g 5.679	s	6.039	6.171	6.286	6.390	6.484	6.657	6.958	7.220
100 (311.0)	v_g 0.01802	$v/10^{-2}$	2.241	2.453	2.639	2.812	2.972	3.275	3.831	4.353
	h_g 2725	h	2926	3017	3097	3172	3241	3373	3624	3868
	s_g 5.615	s	5.947	6.091	6.213	6.321	6.419	6.596	6.902	7.166
110 (318.0)	v_g 0.01598	$v/10^{-2}$	1.960	2.169	2.350	2.514	2.666	2.949	3.465	3.945
	h_g 2705	h	2889	2989	3075	3153	3225	3360	3616	3862
	s_g 5.553	s	5.856	6.014	6.143	6.257	6.358	6.539	6.850	7.117
120 (324.6)	v_g 0.01426	$v/10^{-2}$	1.719	1.931	2.107	2.265	2.410	2.677	3.159	3.605
	h_g 2685	h	2849	2960	3052	3134	3209	3348	3607	3856
	s_g 5.493	s	5.762	5.937	6.076	6.195	6.301	6.487	6.802	7.072
130 (330.8)	v_g 0.01278	$v/10^{-2}$	1.509	1.726	1.901	2.053	2.193	2.447	2.901	3.318
	h_g 2662	h	2804	2929	3028	3114	3192	3335	3599	3850
	s_g 5.433	s	5.664	5.862	6.011	6.136	6.246	6.437	6.758	7.030
140 (336.6)	v_g 0.01149	$v/10^{-2}$	1.321	1.548	1.722	1.872	2.006	2.250	2.679	3.071
	h_g 2638	h	2753	2896	3003	3093	3175	3322	3590	3843
	s_g 5.373	s	5.559	5.784	5.946	6.079	6.193	6.390	6.716	6.991
150 (342.1)	v_g 0.01035	$v/10^{-2}$	1.146	1.391	1.566	1.714	1.844	2.078	2.487	2.857
	h_g 2611	h	2693	2861	2977	3073	3157	3309	3581	3837
	s_g 5.312	s	5.443	5.707	5.883	6.023	6.142	6.345	6.677	6.954
160 (347.3)	v_g 0.00932	$v/10^{-2}$	0.976	1.248	1.427	1.573	1.702	1.928	2.319	2.670
	h_g 2582	h	2617	2821	2949	3051	3139	3295	3573	3831
	s_g 5.248	s	5.304	5.626	5.820	5.968	6.093	6.301	6.639	6.919
170 (352.3)	v_g 0.00838	$v/10^{-2}$		1.117	1.303	1.449	1.576	1.796	2.171	2.506
	h_g 2548	h		2778	2920	3028	3121	3281	3564	3825
	s_g 5.181	s		5.541	5.756	5.914	6.044	6.260	6.603	6.886
180 (357.0)	v_g 0.00751	$v/10^{-2}$		0.997	1.191	1.338	1.463	1.678	2.039	2.359
	h_g 2510	h		2729	2888	3004	3102	3268	3555	3818
	s_g 5.108	s		5.449	5.691	5.861	5.997	6.219	6.569	6.855
190 (361.4)	v_g 0.00668	$v/10^{-2}$		0.882	1.089	1.238	1.362	1.572	1.921	2.228
	h_g 2466	h		2674	2855	2980	3082	3254	3546	3812
	s_g 5.027	s		5.348	5.625	5.807	5.950	6.180	6.536	6.825
200 (365.7)	v_g 0.00585	$v/10^{-2}\,[\text{m}^3/\text{kg}]$		0.768	0.995	1.147	1.270	1.477	1.815	2.110
	h_g 2411	$h/[\text{kJ/kg}]$		2605	2819	2955	3062	3239	3537	3806
	s_g 4.928	$s/[\text{kJ/kg K}]$		5.228	5.556	5.753	5.904	6.142	6.505	6.796
210 (369.8)	v_g 0.00498	$v/10^{-2}$		0.650	0.908	1.064	1.187	1.390	1.719	2.003
	h_g 2336	h		2500	2781	2928	3041	3225	3528	3799
	s_g 4.803	s		5.050	5.484	5.699	5.859	6.105	6.474	6.768
220 (373.7)	v_g 0.00368	$v/10^{-2}$		0.450	0.825	0.987	1.111	1.312	1.632	1.906
	h_g 2178	h		2300	2738	2900	3020	3210	3519	3793
	s_g 4.552	s		4.725	5.409	5.645	5.813	6.068	6.444	6.742
221.2 (374.15)	v_c 0.00317	$v/10^{-2}$	0.163	0.351	0.816	0.978	1.103	1.303	1.622	1.895
	h_c 2084	h	1637	2139	2733	2896	3017	3208	3518	3792
	s_c 4.406	s	3.708	4.490	5.398	5.638	5.807	6.064	6.441	6.739

* See footnote on p. 6.

Note: linear interpolation is not accurate near the critical point.

$\frac{p}{[\text{bar}]}$	$\frac{T}{[°\text{C}]}$	350	375	400	425	450	500	600	700	800
225	$v/10^{-2}[\text{m}^3/\text{kg}]$	0.163	0.249	0.786	0.951	1.076	1.275	1.591	1.861	2.109
	$h/[\text{kJ}/\text{kg}]$	1635	1980	2716	2885	3009	3203	3514	3790	4055
	$s/[\text{kJ}/\text{kg K}]$	3.704	4.470	5.369	5.616	5.790	6.050	6.430	6.729	6.988
250	$v/10^{-2}$	0.160	0.198	0.601	0.789	0.917	1.113	1.412	1.662	1.890
	h	1625	1850	2580	2807	2951	3165	3491	3774	4043
	s	3.682	4.026	5.142	5.474	5.677	5.962	6.361	6.667	6.931
275	$v/10^{-2}$	0.158	0.187	0.419	0.650	0.786	0.980	1.265	1.500	1.710
	h	1617	1814	2382	2718	2890	3125	3468	3758	4032
	s	3.662	3.985	4.828	5.320	5.562	5.878	6.296	6.610	6.878
300	$v/10^{-2}$	0.155	0.180	0.282	0.530	0.674	0.868	1.143	1.364	1.561
	h	1610	1791	2157	2614	2823	3084	3445	3742	4020
	s	3.645	3.933	4.482	5.157	5.444	5.795	6.234	6.557	6.829
350	$v/10^{-2}$	0.152	0.171	0.211	0.343	0.496	0.693	0.952	1.152	1.327
	h	1599	1762	1992	2375	2673	2998	3397	3709	3997
	s	3.614	3.875	4.219	4.776	5.197	5.633	6.120	6.459	6.741
400	$v/10^{-2}$	0.149	0.164	0.191	0.255	0.369	0.562	0.809	0.993	1.152
	h	1590	1743	1935	2203	2514	2906	3348	3677	3974
	s	3.588	3.832	4.119	4.510	4.947	5.474	6.014	6.371	6.662
450	$v/10^{-2}$	0.146	0.160	0.181	0.219	0.291	0.463	0.698	0.870	1.016
	h	1583	1729	1901	2115	2380	2813	3299	3644	3951
	s	3.565	3.797	4.056	4.368	4.740	5.320	5.914	6.290	6.590
500	$v/10^{-2}$	0.144	0.156	0.173	0.201	0.249	0.388	0.611	0.772	0.908
	h	1577	1717	1879	2064	2288	2722	3249	3612	3928
	s	3.544	3.768	4.009	4.279	4.594	5.176	5.821	6.214	6.524
550	$v/10^{-2}$	0.143	0.153	0.168	0.190	0.224	0.334	0.540	0.693	0.820
	h	1572	1709	1862	2030	2227	2641	3200	3579	3905
	s	3.525	3.742	3.971	4.218	4.494	5.047	5.731	6.144	6.462
600	$v/10^{-2}$	0.141	0.151	0.164	0.182	0.209	0.295	0.483	0.627	0.747
	h	1568	1702	1848	2005	2184	2571	3152	3548	3883
	s	3.506	3.718	3.939	4.168	4.419	4.937	5.648	6.077	6.405
650	$v/10^{-2}$	0.139	0.148	0.160	0.176	0.198	0.267	0.436	0.572	0.685
	h	1565	1696	1837	1986	2151	2514	3106	3517	3860
	s	3.489	3.697	3.910	4.128	4.360	4.845	5.568	6.014	6.352
700	$v/10^{-2}$	0.138	0.146	0.157	0.171	0.189	0.247	0.397	0.526	0.633
	h	1561	1691	1829	1971	2127	2468	3062	3486	3839
	s	3.473	3.678	3.886	4.093	4.312	4.769	5.494	5.955	6.300
750	$v/10^{-2}$	0.137	0.145	0.154	0.167	0.183	0.231	0.365	0.486	0.587
	h	1559	1687	1821	1958	2107	2431	3021	3456	3817
	s	3.459	3.659	3.863	4.064	4.272	4.705	5.425	5.899	6.252
800	$v/10^{-2}$	0.136	0.143	0.152	0.163	0.178	0.219	0.338	0.452	0.548
	h	1557	1684	1815	1948	2091	2400	2983	3428	3797
	s	3.444	3.642	3.842	4.037	4.237	4.651	5.361	5.845	6.206
900	$v/10^{-2}$	0.133	0.140	0.148	0.158	0.169	0.202	0.296	0.396	0.484
	h	1554	1678	1805	1932	2066	2353	2916	3373	3756
	s	3.418	3.612	3.805	3.991	4.179	4.563	5.248	5.746	6.120
1000	$v/10^{-2}$	0.131	0.138	0.145	0.153	0.163	0.189	0.267	0.354	0.434
	h	1552	1674	1798	1920	2048	2319	2860	3324	3718
	s	3.394	3.584	3.773	3.951	4.131	4.493	5.153	5.656	6.042

* See footnote on p. 6.

Further Properties of Water and Steam

T [°C]	p_s [bar]	v_f 10^{-2}[m³/kg]	c_{pf}	c_{pg} [kJ/kg K]	μ_f	μ_g 10^{-6}[kg/m s]	k_f	k_g 10^{-6}[kW/m K]	$(Pr)_f$	$(Pr)_g$
0.01	0.006112	0.10002	4.210	1.86	1752	8.49	569	16.3	12.96	0.97
5	0.008719	0.10001	4.204	1.86	1501	8.66	578	16.7	10.92	0.96
10	0.01227	0.10003	4.193	1.86	1300	8.83	587	17.1	9.29	0.96
15	0.01704	0.10010	4.186	1.87	1136	9.00	595	17.5	7.99	0.96
20	0.02337	0.10018	4.183	1.87	1002	9.18	603	17.9	6.95	0.96
25	0.03166	0.10030	4.181	1.88	890	9.35	611	18.3	6.09	0.96
30	0.04242	0.10044	4.179	1.88	797	9.52	618	18.7	5.39	0.96
35	0.05622	0.10060	4.178	1.88	718	9.70	625	19.1	4.80	0.96
40	0.07375	0.10079	4.179	1.89	651	9.87	632	19.5	4.30	0.96
45	0.09582	0.10099	4.181	1.89	594	10.0	638	19.9	3.89	0.95
50	0.1233	0.1012	4.182	1.90	544	10.2	643	20.4	3.54	0.95
55	0.1574	0.1015	4.183	1.90	501	10.4	648	20.8	3.23	0.95
60	0.1992	0.1017	4.185	1.91	463	10.6	653	21.2	2.97	0.95
65	0.2501	0.1020	4.188	1.92	430	10.7	658	21.6	2.74	0.95
70	0.3116	0.1023	4.191	1.93	400	10.9	662	22.0	2.53	0.96
75	0.3855	0.1026	4.194	1.94	374	11.1	666	22.5	2.36	0.96
80	0.4736	0.1029	4.198	1.95	351	11.3	670	22.9	2.20	0.96
85	0.5780	0.1032	4.203	1.96	330	11.4	673	23.3	2.06	0.96
90	0.7011	0.1036	4.208	1.97	311	11.6	676	23.8	1.94	0.96
95	0.8453	0.1040	4.213	1.99	294	11.8	678	24.3	1.83	0.97
100	1.01325	0.1044	4.219	2.01	279	12.0	681	24.8	1.73	0.97
105	1.208	0.1048	4.226	2.03	265	12.2	683	25.3	1.64	0.98
110	1.433	0.1052	4.233	2.05	252	12.4	684	25.8	1.56	0.99
115	1.691	0.1056	4.240	2.07	241	12.6	686	26.3	1.49	0.99
120	1.985	0.1060	4.248	2.09	230	12.8	687	26.8	1.42	1.00
125	2.321	0.1065	4.26	2.12	220	13.0	687	27.3	1.36	1.01
130	2.701	0.1070	4.27	2.15	211	13.2	688	27.8	1.31	1.02
135	3.131	0.1075	4.28	2.18	203	13.4	688	28.3	1.26	1.03
140	3.614	0.1080	4.29	2.21	195	13.5	688	28.8	1.22	1.04
145	4.155	0.1085	4.30	2.25	188	13.7	687	29.4	1.18	1.05
150	4.760	0.1091	4.32	2.29	181	13.9	687	30.0	1.14	1.07
160	6.181	0.1102	4.35	2.38	169	14.2	684	31.3	1.07	1.09
170	7.920	0.1114	4.38	2.49	159	14.6	681	32.6	1.02	1.12
180	10.03	0.1128	4.42	2.62	149	15.0	676	34.1	0.97	1.15
190	12.55	0.1142	4.46	2.76	141	15.3	671	35.7	0.94	1.18
200	15.55	0.1157	4.51	2.91	134	15.7	665	37.5	0.91	1.22
210	19.08	0.1173	4.56	3.07	127	16.0	657	39.4	0.88	1.25
220	23.20	0.1190	4.63	3.25	121	16.3	648	41.5	0.86	1.28
230	27.98	0.1209	4.70	3.45	116	16.7	639	43.9	0.85	1.31
240	33.48	0.1229	4.78	3.68	111	17.1	628	46.5	0.84	1.35
250	39.78	0.1251	4.87	3.94	107	17.5	616	49.5	0.85	1.39
260	46.94	0.1276	4.98	4.22	103	17.9	603	52.8	0.85	1.43
270	55.05	0.1302	5.10	4.55	99	18.3	589	56.6	0.86	1.47
280	64.19	0.1332	5.24	4.98	96	18.8	574	61.0	0.88	1.53
290	74.45	0.1366	5.42	5.46	93	19.3	558	66.0	0.90	1.60
300	85.92	0.1404	5.65	6.18	90	19.8	541	72.0	0.94	1.70
320	112.9	0.1499								
340	146.1	0.1639								
360	186.7	0.1894								
370	210.5	0.2225								
374.15	221.2	0.317								

The values for saturated water can be used with good accuracy above saturation pressure. The values for saturated steam can be used with only moderate accuracy below saturation pressure at temperatures greater than 200 °C.

General Information for H$_2$O

Triple point: Thermodynamic temperature (by definition) =

273.16 K $\cong$ 0.01 °C $\cong$ 491.688 R $\cong$ 32.018 °F

(hence 0 °C $\cong$ 273.15 K, 0 °F $\cong$ 459.67 R, 32 °F $\cong$ 491.67 R)

Gas constant: $R = \tilde{R}/\tilde{m} = 8.3145/18.015 = 0.4615$ kJ/kg K

Compressed Water*

	$T/[°C]$	0.01	100	200	250	300	350	374.15
	p_s	0.006112	1.01325	15.55	39.78	85.92	165.4	221.2
$p/[\text{bar}]$	$v_f/10^{-2}$	0.1000	0.1044	0.1157	0.1251	0.1404	0.1741	0.317
$(T_s/[°C])$	h_f	0	419	852	1086	1345	1671	2084
	s_f	0	1.307	2.331	2.793	3.255	3.779	4.430
100	$(v-v_f)/10^{-2}$	-0.0005	-0.0006	-0.0009	-0.0011	-0.0007		
(311.0)	$(h-h_f)$	$+10$	$+7$	$+4$	0	-2		
	$(s-s_f)$	0.000	-0.008	-0.013	-0.014	-0.007		
221.2	$(v-v_f)/10^{-2}$	-0.0011	-0.0012	-0.0020	-0.0029	-0.0051	-0.0107	0
(374.15)	$(h-h_f)$	$+22$	$+17$	$+9$	$+1$	-12	-34	0
	$(s-s_f)$	$+0.001$	-0.017	-0.031	-0.040	-0.053	-0.071	0
	$(v-v_f)/10^{-2}$	-0.0023	-0.0024	-0.0042	-0.0064	-0.0117	-0.0298	-0.161
500	$(h-h_f)$	$+49$	$+38$	$+23$	$+8$	-21	-94	-369
	$(s-s_f)$	0.000	-0.037	-0.068	-0.091	-0.134	-0.235	-0.670
	$(v-v_f)/10^{-2}$	-0.0044	-0.0044	-0.0075	-0.0111	-0.0191	-0.0427	-0.180
1000	$(h-h_f)$	$+96$	$+76$	$+51$	$+28$	-17	-119	-415
	$(s-s_f)$	-0.007	-0.070	-0.124	-0.164	-0.235	-0.385	-0.853

* See footnote on p. 6.

Saturated Ice and Steam

T	p_s	v_i	v_g	u_i	u_g	h_i	h_g	s_i	s_g
[°C]	[bar]	$10^{-2}[\text{m}^3/\text{kg}]$	$[\text{m}^3/\text{kg}]$	[kJ/kg]		[kJ/kg]		[kJ/kg K]	
0.01	0.006112	0.1091	206.1	-333.5	2374.7	-333.5	2500.8	-1.221	9.155
-10	0.002598	0.1089	467.5	-354.2	2360.8	-354.2	2482.2	-1.298	9.481
-20	0.001038	0.1087	1125	-374.1	2346.8	-374.1	2463.6	-1.375	9.835
-30	0.0003809	0.1086	2946	-393.3	2332.9	-393.3	2445.1	-1.452	10.221
-40	0.0001288	0.1084	8354	-411.8	2319.0	-411.8	2426.6	-1.530	10.644

Isentropic Expansion of Steam—Approximate Relations

Wet equilibrium expansion:

pv^n = constant, with $n \approx 1.035 + 0.1x_1$ for steam with an initial dryness fraction $0.7 < x_1 < 1.0$

Superheated and supersaturated expansion:

pv^n = constant and $p/T^{n/(n-1)}$ = constant, with $n \approx 1.3$

Enthalpy drop $\dfrac{(h_2 - h_1)}{[\text{kJ/kg}]} = \left(\dfrac{h_1}{[\text{kJ/kg}]} - 1943 \right) \left[\left(\dfrac{p_2}{p_1} \right)^{(n-1)/n} - 1 \right]$

Specific volume of supersaturated steam:

$\dfrac{p}{[\text{bar}]} \times \dfrac{v}{[\text{m}^3/\text{kg}]} \times 10^2 = \dfrac{0.3}{1.3} \left(\dfrac{h}{[\text{kJ/kg}]} - 1943 \right)$

11

Mercury – Hg

p [bar]	T_s [°C]	v_g [m³/kg]	h_f	h_{fg} [kJ/kg]	h_g	s_f	s_{fg} [kJ/kg K]	s_g
0.0006	109.2	259.6	15.13	297.20	312.33	0.0466	0.7774	0.8240
0.0007	112.3	224.3	15.55	297.14	312.69	0.0477	0.7709	0.8186
0.0008	115.0	197.7	15.93	297.09	313.02	0.0487	0.7654	0.8141
0.0009	117.5	176.8	16.27	297.04	313.31	0.0496	0.7604	0.8100
0.0010	119.7	160.1	16.58	297.00	313.58	0.0503	0.7560	0.8063
0.002	134.9	83.18	18.67	296.71	315.38	0.0556	0.7271	0.7827
0.004	151.5	43.29	20.93	296.40	317.33	0.0610	0.6981	0.7591
0.006	161.8	29.57	22.33	296.21	318.54	0.0643	0.6811	0.7454
0.008	169.4	22.57	23.37	296.06	319.43	0.0666	0.6690	0.7356
0.010	175.5	18.31	24.21	295.95	320.16	0.0685	0.6596	0.7281
0.02	195.6	9.570	26.94	295.57	322.51	0.0744	0.6305	0.7049
0.04	217.7	5.013	29.92	295.15	325.07	0.0806	0.6013	0.6819
0.06	231.6	3.438	31.81	294.89	326.70	0.0843	0.5842	0.6685
0.08	242.0	2.632	33.21	294.70	327.91	0.0870	0.5721	0.6591
0.10	250.3	2.140	34.33	294.54	328.87	0.0892	0.5627	0.6519
0.2	278.1	1.128	38.05	294.02	332.07	0.0961	0.5334	0.6295
0.4	309.1	0.5942	42.21	293.43	335.64	0.1034	0.5039	0.6073
0.6	329.0	0.4113	44.85	293.06	337.91	0.1078	0.4869	0.5947
0.8	343.9	0.3163	46.84	292.78	339.62	0.1110	0.4745	0.5855
1	356.1	0.2581	48.45	292.55	341.00	0.1136	0.4649	0.5785
2	397.1	0.1377	53.87	291.77	345.64	0.1218	0.4353	0.5571
3	423.8	0.09551	57.38	291.27	348.65	0.1268	0.4179	0.5447
4	444.1	0.07378	60.03	290.89	350.92	0.1305	0.4056	0.5361
5	460.7	0.06044	62.20	290.58	352.78	0.1334	0.3960	0.5294
6	474.9	0.05137	64.06	290.31	354.37	0.1359	0.3881	0.5240
7	487.3	0.04479	65.66	290.08	355.74	0.1380	0.3815	0.5195
8	498.4	0.03978	67.11	289.87	356.98	0.1398	0.3757	0.5155
9	508.5	0.03584	68.42	289.68	358.10	0.1415	0.3706	0.5121
10	517.8	0.03266	69.61	289.50	359.11	0.1429	0.3660	0.5089
12	534.4	0.02781	71.75	289.19	360.94	0.1455	0.3581	0.5036
14	549.0	0.02429	73.63	288.92	362.55	0.1478	0.3514	0.4992
16	562.0	0.02161	75.37	288.67	364.04	0.1498	0.3456	0.4954
18	574.0	0.01949	76.83	288.45	365.28	0.1515	0.3405	0.4920
20	584.9	0.01778	78.23	288.24	366.47	0.1531	0.3359	0.4890
22	595.1	0.01637	79.54	288.05	367.59	0.1546	0.3318	0.4864
24	604.6	0.01518	80.75	287.87	368.62	0.1559	0.3280	0.4839
26	613.5	0.01416	81.89	287.70	369.59	0.1571	0.3245	0.4816
28	622.0	0.01329	82.96	287.54	370.50	0.1583	0.3212	0.4795
30	630.0	0.01252	83.97	287.39	371.36	0.1594	0.3182	0.4776
35	648.5	0.01096	86.33	287.04	373.37	0.1619	0.3115	0.4734
40	665.1	0.00978	88.43	286.73	375.16	0.1641	0.3056	0.4697
45	680.3	0.00885	90.35	286.44	376.79	0.1660	0.3004	0.4664
50	694.4	0.00809	92.11	286.18	378.29	0.1678	0.2958	0.4636
55	707.4	0.00746	93.76	285.93	379.69	0.1694	0.2916	0.4610
60	719.7	0.00693	95.30	285.70	381.00	0.1709	0.2878	0.4587
65	731.3	0.00648	96.75	285.48	382.23	0.1723	0.2842	0.4565
70	742.3	0.00609	98.12	285.28	383.40	0.1736	0.2809	0.4545
75	752.7	0.00575	99.42	285.08	384.50	0.1748	0.2779	0.4527

h_f and s_f are zero at 0 °C. Molar mass $\tilde{m} = 200.59$ kg/kmol; for superheated vapour $c_p = 0.1036$ kJ/kg K; further properties of the liquid are given on p. 23.

Ammonia – NH$_3$ (Refrigerant 717)

		Saturation Values						Superheat ($T-T_s$)		
								50 K		100 K
T	p_s	v_g	h_f	h_g	s_f	s_g		h	s	h	s
[°C]	[bar]	[m^3/kg]	[kJ/kg]		[kJ/kg K]			[kJ/kg]	[kJ/kg K]	[kJ/kg]	[kJ/kg K]
−50	0.4089	2.625	−44.4	1373.3	−0.194	6.159		1479.8	6.592	1585.9	6.948
−45	0.5454	2.005	−22.3	1381.6	−0.096	6.057		1489.3	6.486	1596.1	6.839
−40	0.7177	1.552	0	1390.0	0	5.962		1498.6	6.387	1606.3	6.736
−35	0.9322	1.216	22.3	1397.9	0.095	5.872		1507.9	6.293	1616.3	6.639
−30	1.196	0.9633	44.7	1405.6	0.188	5.785		1517.0	6.203	1626.3	6.547
−28	1.317	0.8809	53.6	1408.5	0.224	5.751		1520.7	6.169	1630.3	6.512
−26	1.447	0.8058	62.6	1411.4	0.261	5.718		1524.3	6.135	1634.2	6.477
−24	1.588	0.7389	71.7	1414.3	0.297	5.686		1527.9	6.103	1638.2	6.444
−22	1.740	0.6783	80.8	1417.3	0.333	5.655		1531.4	6.071	1642.2	6.411
−20	1.902	0.6237	89.8	1420.0	0.368	5.623		1534.8	6.039	1646.0	6.379
−18	2.077	0.5743	98.8	1422.7	0.404	5.593		1538.2	6.008	1650.0	6.347
−16	2.265	0.5296	107.9	1425.3	0.440	5.563		1541.7	5.978	1653.8	6.316
−14	2.465	0.4890	117.0	1427.9	0.475	5.533		1545.1	5.948	1657.7	6.286
−12	2.680	0.4521	126.2	1430.5	0.510	5.504		1548.5	5.919	1661.5	6.256
−10	2.908	0.4185	135.4	1433.0	0.544	5.475		1551.7	5.891	1665.3	6.227
−8	3.153	0.3879	144.5	1435.3	0.579	5.447		1554.9	5.863	1669.0	6.199
−6	3.413	0.3599	153.6	1437.6	0.613	5.419		1558.2	5.836	1672.8	6.171
−4	3.691	0.3344	162.8	1439.9	0.647	5.392		1561.4	5.808	1676.4	6.143
−2	3.983	0.3110	172.0	1442.2	0.681	5.365		1564.6	5.782	1680.1	6.116
0	4.295	0.2895	181.2	1444.4	0.715	5.340		1567.8	5.756	1683.9	6.090
2	4.625	0.2699	190.4	1446.5	0.749	5.314		1570.9	5.731	1687.5	6.065
4	4.975	0.2517	199.7	1448.5	0.782	5.288		1574.0	5.706	1691.2	6.040
6	5.346	0.2351	209.1	1450.6	0.816	5.263		1577.0	5.682	1694.9	6.015
8	5.736	0.2198	218.5	1452.5	0.849	5.238		1580.1	5.658	1698.4	5.991
10	6.149	0.2056	227.8	1454.3	0.881	5.213		1583.1	5.634	1702.2	5.967
12	6.585	0.1926	237.2	1456.1	0.914	5.189		1586.0	5.611	1705.7	5.943
14	7.045	0.1805	246.6	1457.8	0.947	5.165		1588.9	5.588	1709.1	5.920
16	7.529	0.1693	256.0	1459.5	0.979	5.141		1591.7	5.565	1712.5	5.898
18	8.035	0.1590	265.5	1461.1	1.012	5.118		1594.4	5.543	1715.9	5.876
20	8.570	0.1494	275.1	1462.6	1.044	5.095		1597.2	5.521	1719.3	5.854
22	9.134	0.1405	284.6	1463.9	1.076	5.072		1600.0	5.499	1722.8	5.832
24	9.722	0.1322	294.1	1465.2	1.108	5.049		1602.7	5.478	1726.3	5.811
26	10.34	0.1245	303.7	1466.5	1.140	5.027		1605.3	5.458	1729.6	5.790
28	10.99	0.1173	313.4	1467.8	1.172	5.005		1608.0	5.437	1732.7	5.770
30	11.67	0.1106	323.1	1468.9	1.204	4.984		1610.5	5.417	1735.9	5.750
32	12.37	0.1044	332.8	1469.9	1.235	4.962		1613.0	5.397	1739.3	5.731
34	13.11	0.0986	342.5	1470.8	1.267	4.940		1615.4	5.378	1742.6	5.711
36	13.89	0.0931	352.3	1471.8	1.298	4.919		1617.8	5.358	1745.7	5.692
38	14.70	0.0880	362.1	1472.6	1.329	4.898		1620.1	5.340	1748.7	5.674
40	15.54	0.0833	371.9	1473.3	1.360	4.877		1622.4	5.321	1751.9	5.655
42	16.42	0.0788	381.8	1473.8	1.391	4.856		1624.6	5.302	1755.0	5.637
44	17.34	0.0746	391.8	1474.2	1.422	4.835		1626.8	5.284	1758.0	5.619
46	18.30	0.0706	401.8	1474.5	1.453	4.814		1629.0	5.266	1761.0	5.602
48	19.29	0.0670	411.9	1474.7	1.484	4.793		1631.1	5.248	1764.0	5.584
50	20.33	0.0635	421.9	1474.7	1.515	4.773		1633.1	5.230	1766.8	5.567

Critical point $T_c = 132.4$ °C, $p_c = 113.0$ bar.
Molar mass $\bar{m} = 17.030$ kg/kmol; further properties of the liquid are given on p. 23.

Dichlorodifluoromethane – CF_2Cl_2 (Refrigerant 12)

							Superheat ($T-T_s$)			
		Saturation Values					15 K		30 K	
T	p_s	v_g	h_f	h_g	s_f	s_g	h	s	h	s
[°C]	[bar]	[m³/kg]	[kJ/kg]		[kJ/kg K]		[kJ/kg]	[kJ/kg K]	[kJ/kg]	[kJ/kg K]
−100	0.0118	10.100	−51.84	142.00	−0.2567	0.8628	148.89	0.9019	156.10	0.9428
−95	0.0181	6.585	−47.56	144.22	−0.2323	0.8442	151.23	0.8830	158.55	0.9195
−90	0.0284	4.416	−43.28	146.46	−0.2086	0.8274	153.59	0.8649	161.02	0.9010
−85	0.0424	3.037	−39.00	148.73	−0.1856	0.8122	155.98	0.8493	163.52	0.8851
−80	0.0617	2.138	−34.72	151.02	−0.1631	0.7985	158.39	0.8351	166.04	0.8706
−75	0.0879	1.538	−30.43	153.32	−0.1412	0.7861	160.82	0.8226	168.57	0.8578
−70	0.1227	1.127	−26.13	155.63	−0.1198	0.7749	163.26	0.8110	171.12	0.8459
−65	0.1680	0.8412	−21.81	157.96	−0.0988	0.7649	165.70	0.8008	173.68	0.8355
−60	0.2262	0.6379	−17.49	160.29	−0.0783	0.7558	168.15	0.7915	176.26	0.8259
−55	0.2998	0.4910	−13.14	162.62	−0.0582	0.7475	170.60	0.7830	178.84	0.8172
−50	0.3915	0.3831	−8.78	164.95	−0.0384	0.7401	173.07	0.7753	181.43	0.8093
−45	0.5044	0.3027	−4.40	167.28	−0.0190	0.7335	175.54	0.7685	184.01	0.8023
−40	0.6417	0.2419	0	169.60	0	0.7274	178.00	0.7623	186.60	0.7959
−35	0.8071	0.1954	4.42	171.90	0.0187	0.7219	180.45	0.7568	189.18	0.7902
−30	1.004	0.1594	8.86	174.20	0.0371	0.7170	182.90	0.7517	191.76	0.7851
−25	1.237	0.1312	13.33	176.48	0.0552	0.7127	185.33	0.7473	194.33	0.7805
−20	1.509	0.1088	17.82	178.73	0.0731	0.7087	187.75	0.7432	196.89	0.7764
−15	1.826	0.0910	22.33	180.97	0.0906	0.7051	190.15	0.7397	199.44	0.7728
−10	2.191	0.0766	26.87	183.19	0.1080	0.7020	192.53	0.7365	201.97	0.7695
−5	2.610	0.0650	31.45	185.38	0.1251	0.6991	194.90	0.7336	204.49	0.7666
0	3.086	0.0554	36.05	187.53	0.1420	0.6966	197.25	0.7311	206.99	0.7641
5	3.626	0.0475	40.69	189.66	0.1587	0.6943	199.56	0.7289	209.47	0.7618
10	4.233	0.0409	45.37	191.74	0.1752	0.6921	201.85	0.7268	211.92	0.7598
15	4.914	0.0354	50.10	193.78	0.1915	0.6901	204.10	0.7251	214.35	0.7580
20	5.673	0.0308	54.87	195.78	0.2078	0.6885	206.32	0.7235	216.75	0.7565
25	6.516	0.0269	59.70	197.73	0.2239	0.6869	208.50	0.7220	219.11	0.7552
30	7.449	0.0235	64.59	199.62	0.2399	0.6853	210.63	0.7208	221.44	0.7540
35	8.477	0.0206	69.55	201.45	0.2559	0.6839	212.72	0.7196	223.73	0.7529
40	9.607	0.0182	74.59	203.20	0.2718	0.6825	214.76	0.7185	225.98	0.7519
45	10.84	0.0160	79.71	204.87	0.2877	0.6811	216.74	0.7175	228.18	0.7511
50	12.19	0.0142	84.94	206.45	0.3037	0.6797	218.64	0.7166	230.33	0.7503
55	13.66	0.0125	90.27	207.92	0.3197	0.6782	220.48	0.7156	232.42	0.7496
60	15.26	0.0111	95.74	209.26	0.3358	0.6765	222.23	0.7146	234.45	0.7490
65	16.99	0.00985	101.36	210.46	0.3521	0.6747	223.89	0.7136	236.42	0.7484
70	18.86	0.00873	107.15	211.48	0.3686	0.6726	225.45	0.7125	238.32	0.7477
75	20.88	0.00772	113.15	212.29	0.3854	0.6702	226.89	0.7113	240.13	0.7470
80	23.05	0.00682	119.39	212.83	0.4027	0.6673	228.21	0.7099	241.86	0.7463
85	25.38	0.00601	125.93	213.04	0.4204	0.6636	229.39	0.7084	243.50	0.7455
90	27.89	0.00526	132.84	212.80	0.4389	0.6591	230.43	0.7067	245.03	0.7445
95	30.57	0.00456	140.23	211.94	0.4583	0.6531	231.30	0.7047	246.47	0.7435
100	33.44	0.00390	148.32	210.12	0.4793	0.6449	231.93	0.7023	247.80	0.7424
105	36.51	0.00324	157.52	206.57	0.5028	0.6325	232.22	0.6994	248.97	0.7412
110	39.79	0.00246	169.55	197.99	0.5334	0.6076	232.47	0.6964	250.10	0.7399
112	41.15	0.00179	183.43	183.43	0.5690	0.5690	232.80	0.6958	250.58	0.7394

Molar mass $\tilde{m} = 120.91$ kg/kmol; further properties of the liquid are given on p. 23.

Tetrafluoroethane – CH_2F–CF_3 (Refrigerant 134a)

							Superheat $(T - T_s)$			
Saturation Values							10 K		20 K	
T	p_s	v_g	h_f	h_g	s_f	s_g	h	s	h	s
[°C]	[bar]	[m³/kg]	[kJ/kg]		[kJ/kg K]		[kJ/kg]	[kJ/kg K]	[kJ/kg]	[kJ/kg K]
−103.30	0.0041	34.032	77.69	335.24	0.4453	1.9616	341.16	1.9955	347.29	2.0287
−100	0.0058	24.341	80.89	337.15	0.4640	1.9439	343.14	1.9776	349.35	2.0106
− 90	0.0155	9.5984	90.97	343.05	0.5205	1.8969	349.27	1.9300	355.70	1.9624
− 80	0.0370	4.2333	101.60	349.09	0.5770	1.8584	355.55	1.8910	362.20	1.9229
− 70	0.0800	2.0522	112.70	355.25	0.6330	1.8270	361.95	1.8592	368.84	1.8907
− 60	0.1591	1.07785	124.23	361.48	0.6884	1.8015	368.44	1.8334	375.57	1.8646
− 50	0.2944	0.60592	136.14	367.76	0.7430	1.7809	374.99	1.8126	382.38	1.8436
− 40	0.5188	0.36089	148.37	374.03	0.7965	1.7644	381.56	1.7960	389.22	1.8269
− 30	0.8435	0.22577	160.89	380.27	0.8490	1.7512	388.12	1.7828	396.07	1.8137
− 25	1.0637	0.18146	167.25	383.37	0.8748	1.7457	391.38	1.7774	399.49	1.8082
− 20	1.3272	0.14725	173.67	386.44	0.9003	1.7408	394.63	1.7726	402.90	1.8034
− 15	1.6393	0.12055	180.16	389.49	0.9256	1.7365	397.86	1.7683	406.29	1.7992
− 10	2.0060	0.09949	186.71	392.51	0.9506	1.7327	401.07	1.7647	409.67	1.7956
− 5	2.4335	0.08273	193.32	395.49	0.9754	1.7294	404.25	1.7614	413.02	1.7924
0†	2.9281	0.06925	200.00†	398.43	1.0000†	1.7264	407.40	1.7587	416.35	1.7897
5	3.4966	0.05834	206.75	401.33	1.0243	1.7238	410.50	1.7562	419.65	1.7874
10	4.1459	0.04942	213.57	404.16	1.0484	1.7215	413.56	1.7542	422.90	1.7855
15	4.8833	0.04208	220.46	406.93	1.0723	1.7194	416.57	1.7524	426.12	1.7838
20	5.7162	0.03599	227.45	409.62	1.0961	1.7176	419.52	1.7508	429.29	1.7825
25	6.6525	0.03092	234.52	412.23	1.1198	1.7158	422.41	1.7494	432.40	1.7813
30	7.7000	0.02665	241.69	414.74	1.1434	1.7142	425.21	1.7482	435.44	1.7803
35	8.8672	0.02304	248.98	417.14	1.1669	1.7126	427.93	1.7470	438.42	1.7795
40	10.163	0.01998	256.38	419.41	1.1903	1.7109	430.55	1.7460	441.32	1.7788
45	11.595	0.01735	263.92	421.53	1.2138	1.7092	433.06	1.7449	444.13	1.7781
50	13.174	0.01510	271.61	423.47	1.2374	1.7073	435.44	1.7438	446.84	1.7775
55	14.910	0.01315	279.46	425.20	1.2610	1.7051	437.69	1.7426	449.45	1.7769
60	16.812	0.01145	287.51	426.69	1.2848	1.7026	439.77	1.7412	451.93	1.7762
65	18.892	0.00997	295.77	427.89	1.3088	1.6995	441.67	1.7397	454.29	1.7754
70	21.161	0.00866	304.29	428.72	1.3332	1.6958	443.36	1.7378	456.50	1.7745
75	23.633	0.00750	313.13	429.09	1.3580	1.6911	444.82	1.7356	458.54	1.7734
80	26.323	0.00645	322.36	428.85	1.3835	1.6851	446.01	1.7330	460.42	1.7721
85	29.249	0.00550	332.16	427.77	1.4101	1.6771	446.88	1.7298	462.09	1.7706
90	32.433	0.00462	342.79	425.40	1.4386	1.6661	447.40	1.7259	463.55	1.7687
95	35.906	0.00375	355.05	420.64	1.4709	1.6491	447.49	1.7212	464.76	1.7663
100	39.728	0.00266	373.53	406.93	1.5193	1.6088	447.04	1.7153	465.65	1.7633
101.00	40.550	0.00196	389.67	389.67	1.5621	1.5621	446.84	1.7139	465.77	1.7626

Molar mass $\bar{m}$ = 102.03 kg/kmol; further properties of the liquid are given on p. 23.

†The datum state for refrigerant properties used to be −40 °C (h_f = 0, s_f = 0), a temperature at which −40 °C = −40 °F. This datum state is used here for the R717 and R12 tables. Nowadays the datum state chosen is 0 °C (h_f = 200 kJ/kg, s_f = 1.000 kJ/kg K), a choice which ensures that no negative values of h_f and s_f appear in common refrigerant tables. This datum state is chosen for the R134a table.

It must be remembered that datum states are quite arbitrary and do not affect calculations which involve changes of properties, such as Δh.

Dry Air at Low Pressure

							at 1 atm	
T	c_p	c_v	γ	μ	k	Pr	ρ	v
[K]	[kJ/kg K]			10^{-5}[kg/m s]	10^{-5}[kW/m K]		[kg/m³]	10^{-5}[m²/s]
175	1.0023	0.7152	1.401	1.182	1.593	0.744	2.017	0.586
200	1.0025	0.7154	1.401	1.329	1.809	0.736	1.765	0.753
225	1.0027	0.7156	1.401	1.467	2.020	0.728	1.569	0.935
250	1.0031	0.7160	1.401	1.599	2.227	0.720	1.412	1.132
275	1.0038	0.7167	1.401	1.725	2.428	0.713	1.284	1.343
300	1.0049	0.7178	1.400	1.846	2.624	0.707	1.177	1.568
325	1.0063	0.7192	1.400	1.962	2.816	0.701	1.086	1.807
350	1.0082	0.7211	1.398	2.075	3.003	0.697	1.009	2.056
375	1.0106	0.7235	1.397	2.181	3.186	0.692	0.9413	2.317
400	1.0135	0.7264	1.395	2.286	3.365	0.688	0.8824	2.591
450	1.0206	0.7335	1.391	2.485	3.710	0.684	0.7844	3.168
500	1.0295	0.7424	1.387	2.670	4.041	0.680	0.7060	3.782
550	1.0398	0.7527	1.381	2.849	4.357	0.680	0.6418	4.439
600	1.0511	0.7640	1.376	3.017	4.661	0.680	0.5883	5.128
650	1.0629	0.7758	1.370	3.178	4.954	0.682	0.5430	5.853
700	1.0750	0.7879	1.364	3.332	5.236	0.684	0.5043	6.607
750	1.0870	0.7999	1.359	3.482	5.509	0.687	0.4706	7.399
800	1.0987	0.8116	1.354	3.624	5.774	0.690	0.4412	8.214
850	1.1101	0.8230	1.349	3.763	6.030	0.693	0.4153	9.061
900	1.1209	0.8338	1.344	3.897	6.276	0.696	0.3922	9.936
950	1.1313	0.8442	1.340	4.026	6.520	0.699	0.3716	10.83
1000	1.1411	0.8540	1.336	4.153	6.754	0.702	0.3530	11.76
1050	1.1502	0.8631	1.333	4.276	6.985	0.704	0.3362	12.72
1100	1.1589	0.8718	1.329	4.396	7.209	0.707	0.3209	13.70
1150	1.1670	0.8799	1.326	4.511	7.427	0.709	0.3069	14.70
1200	1.1746	0.8875	1.323	4.626	7.640	0.711	0.2941	15.73
1250	1.1817	0.8946	1.321	4.736	7.849	0.713	0.2824	16.77
1300	1.1884	0.9013	1.319	4.846	8.054	0.715	0.2715	17.85
1350	1.1946	0.9075	1.316	4.952	8.253	0.717	0.2615	18.94
1400	1.2005	0.9134	1.314	5.057	8.450	0.719	0.2521	20.06
1500	1.2112	0.9241	1.311	5.264	8.831	0.722	0.2353	22.36
1600	1.2207	0.9336	1.308	5.457	9.199	0.724	0.2206	24.74
1700	1.2293	0.9422	1.305	5.646	9.554	0.726	0.2076	27.20
1800	1.2370	0.9499	1.302	5.829	9.899	0.728	0.1961	29.72
1900	1.2440	0.9569	1.300	6.008	10.233	0.730	0.1858	32.34
2000	1.2505	0.9634	1.298	—	—	—	0.1765	—
2100	1.2564	0.9693	1.296	—	—	—	0.1681	—
2200	1.2619	0.9748	1.295	—	—	—	0.1604	—
2300	1.2669	0.9798	1.293	—	—	—	0.1535	—
2400	1.2717	0.9846	1.292	—	—	—	0.1471	—
2500	1.2762	0.9891	1.290	—	—	—	0.1412	—
2600	1.2803	0.9932	1.289	—	—	—	0.1358	—
2700	1.2843	0.9972	1.288	—	—	—	0.1307	—
2800	1.2881	1.0010	1.287	—	—	—	0.1261	—
2900	1.2916	1.0045	1.286	—	—	—	0.1217	—
3000	1.2949	1.0078	1.285	—	—	—	0.1177	—

The values for air can also be used with reasonable accuracy for CO, N_2 and O_2.

The values of the thermodynamic properties c_v and c_p on pp. 16 and 17 are those at zero pressure. The values for the gases are quite accurate over a wide range of pressure, but those for the vapours increase appreciably with pressure.

The transport properties μ and k for air are accurate over a wide range of pressure, except at such low pressures that the mean free path of the molecules is comparable to the distance between the solid surfaces containing the gas.

At high temperatures (> 1500 K for air) dissociation becomes appreciable and pressure is a significant variable for both gases and vapours: the values on pp. 16 and 17 apply only to undissociated states.

Specific Heat Capacity c_p/[kJ/kg K] of Some Gases and Vapours

T/[K]	CO_2	CO	H_2	N_2	O_2	H_2O	CH_4	C_2H_4	C_2H_6
175	0.709	1.039	13.12	1.039	0.910	1.850	2.083	1.241	
200	0.735	1.039	13.53	1.039	0.910	1.851	2.087	1.260	
225	0.763	1.039	13.83	1.039	0.911	1.852	2.121	1.316	
250	0.791	1.039	14.05	1.039	0.913	1.855	2.156	1.380	1.535
275	0.819	1.040	14.20	1.039	0.915	1.859	2.191	1.453	1.651
300	0.846	1.040	14.31	1.040	0.918	1.864	2.226	1.535	1.766
325	0.871	1.041	14.38	1.040	0.923	1.871	2.293	1.621	1.878
350	0.895	1.043	14.43	1.041	0.928	1.880	2.365	1.709	1.987
375	0.918	1.045	14.46	1.042	0.934	1.890	2.442	1.799	2.095
400	0.939	1.048	14.48	1.044	0.941	1.901	2.525	1.891	2.199
450	0.978	1.054	14.50	1.049	0.956	1.926	2.703	2.063	2.402
500	1.014	1.064	14.51	1.056	0.972	1.954	2.889	2.227	2.596
550	1.046	1.075	14.53	1.065	0.988	1.984	3.074	2.378	2.782
600	1.075	1.087	14.55	1.075	1.003	2.015	3.256	2.519	2.958
650	1.102	1.100	14.57	1.086	1.017	2.047	3.432	2.649	3.126
700	1.126	1.113	14.60	1.098	1.031	2.080	3.602	2.770	3.286
750	1.148	1.126	14.65	1.110	1.043	2.113	3.766	2.883	3.438
800	1.168	1.139	14.71	1.122	1.054	2.147	3.923	2.989	3.581
850	1.187	1.151	14.77	1.134	1.065	2.182	4.072	3.088	3.717
900	1.204	1.163	14.83	1.146	1.074	2.217	4.214	3.180	3.846
950	1.220	1.174	14.90	1.157	1.082	2.252	4.348	3.266	
1000	1.234	1.185	14.98	1.167	1.090	2.288	4.475	3.347	
1050	1.247	1.194	15.06	1.177	1.097	2.323	4.595	3.423	
1100	1.259	1.203	15.15	1.187	1.103	2.358	4.708	3.494	
1150	1.270	1.212	15.25	1.196	1.109	2.392	4.814	3.561	
1200	1.280	1.220	15.34	1.204	1.115	2.425			
1250	1.290	1.227	15.44	1.212	1.120	2.458			
1300	1.298	1.234	15.54	1.219	1.125	2.490			
1350	1.306	1.240	15.65	1.226	1.130	2.521			
1400	1.313	1.246	15.77	1.232	1.134	2.552			
1500	1.326	1.257	16.02	1.244	1.143	2.609			
1600	1.338	1.267	16.23	1.254	1.151	2.662			
1700	1.348	1.275	16.44	1.263	1.158	2.711			
1800	1.356	1.282	16.64	1.271	1.166	2.756			
1900	1.364	1.288	16.83	1.278	1.173	2.798			
2000	1.371	1.294	17.01	1.284	1.181	2.836			
2100	1.377	1.299	17.18	1.290	1.188	2.872			
2200	1.383	1.304	17.35	1.295	1.195	2.904			
2300	1.388	1.308	17.50	1.300	1.202	2.934			
2400	1.393	1.311	17.65	1.304	1.209	2.962			
2500	1.397	1.315	17.80	1.307	1.216	2.987			
2600	1.401	1.318	17.93	1.311	1.223	3.011			
2700	1.404	1.321	18.06	1.314	1.230	3.033			
2800	1.408	1.324	18.17	1.317	1.236	3.053			
2900	1.411	1.326	18.28	1.320	1.243	3.072			
3000	1.414	1.329	18.39	1.323	1.249	3.090			
3500	1.427	1.339	18.91	1.333	1.276	3.163			
4000	1.437	1.346	19.39	1.342	1.299	3.217			
4500	1.446	1.353	19.83	1.349	1.316	3.258			
5000	1.455	1.359	20.23	1.355	1.328	3.292			
5500	1.465	1.365	20.61	1.362	1.337	3.322			
6000	1.476	1.370	20.96	1.369	1.344	3.350			

T/[K]	C_6H_6	C_8H_{18}
250	0.850	1.308
275	0.957	1.484
300	1.060	1.656
325	1.160	1.825
350	1.255	1.979
375	1.347	2.109
400	1.435	2.218
450	1.600	2.403
500	1.752	2.608
550	1.891	2.774
600	2.018	2.924
650	2.134	3.121
700	2.239	3.232
750	2.335	3.349
800	2.422	3.465
850	2.500	3.582
900	2.571	3.673

The specific heat capacities of atomic H, N and O are given with adequate accuracy by $c_p = 2.5\, \tilde{R}/\tilde{m}$ where $\tilde{m}$ is the molar mass of the *atomic* species.

Molar Properties of Some Gases and Vapours

By definition: $\tilde{h} = \tilde{u} + p\tilde{v}$ and $\tilde{g} = \tilde{h} - T\tilde{s}$

$\tilde{h}$ and $\tilde{u}$ are virtually independent of pressure and in the following will be treated as such: hence

$$\tilde{u} = \tilde{h} - \tilde{R}T$$

$\tilde{s}$ and $\tilde{g}$ are tabulated for states at the standard pressure $p^{\circ} = 1$ bar and are denoted by $\tilde{s}^{\circ}$ and $\tilde{g}^{\circ}$. At any other pressure p, $\tilde{s}$ and $\tilde{g}$ at a given temperature T can be found from

$$\tilde{s} - \tilde{s}^{\circ} = -\tilde{R}\ln(p/p^{\circ})$$

$$\tilde{g} - \tilde{g}^{\circ} = (\tilde{h} - \tilde{h}^{\circ}) - T(\tilde{s} - \tilde{s}^{\circ}) = +\tilde{R}T\ln(p/p^{\circ})$$

For individual gases and vapours, changes in $\tilde{s}$ and $\tilde{g}$ between states (p_1, T_1) and (p_2, T_2) are given by

$$\tilde{s}_2 - \tilde{s}_1 = (\tilde{s}_2 - \tilde{s}_2^{\circ}) + (\tilde{s}_2^{\circ} - \tilde{s}_1^{\circ}) + (\tilde{s}_1^{\circ} - \tilde{s}_1)$$
$$= (\tilde{s}_2^{\circ} - \tilde{s}_1^{\circ}) - \tilde{R}\ln(p_2/p_1)$$

$$\tilde{g}_2 - \tilde{g}_1 = (\tilde{g}_2 - \tilde{g}_2^{\circ}) + (\tilde{g}_2^{\circ} - \tilde{g}_1^{\circ}) + (\tilde{g}_1^{\circ} - \tilde{g}_1)$$
$$= (\tilde{g}_2^{\circ} - \tilde{g}_1^{\circ}) + \tilde{R}T_2\ln(p_2/p^{\circ}) - \tilde{R}T_1\ln(p_1/p^{\circ})$$

For a constituent in a mixture, p_1 and p_2 must be regarded as the partial pressures in the respective states.

When performing calculations involving non-reacting mixtures, the datum states at which $\tilde{h}$ and $\tilde{s}$ are arbitrarily put equal to zero are unimportant: in the following tables they are (1 bar, 298.15 K) for $\tilde{h}$ and (1 bar, 0.0 K) for $\tilde{s}$. The datum states are important when chemical reactions are involved—see p. 20.

$\tilde{h}$	$\tilde{u}$	$\tilde{s}^{\circ}$	$\tilde{g}^{\circ}$	T	$\tilde{h}$	$\tilde{u}$	$\tilde{s}^{\circ}$	$\tilde{g}^{\circ}$
[kJ/kmol]	[kJ/kmol]	[kJ/kmol K]	[kJ/kmol]	[K]	[kJ/kmol]	[kJ/kmol]	[kJ/kmol K]	[kJ/kmol]
Carbon Dioxide (CO_2)		$\tilde{m} = 44.010\,\dfrac{kg}{kmol}$			Water Vapour (H_2O)		$\tilde{m} = 18.015\,\dfrac{kg}{kmol}$	
−9 364	−9 364	0	− 9 364	0	−9 904	−9 904	0	− 9 904
−6 456	−7 287	178.90	− 24 346	100	−6 615	−7 446	152.28	− 21 843
−3 414	−5 077	199.87	− 43 387	200	−3 280	−4 943	175.38	− 38.356
0	−2 479	213.69	− 63 710	298.15	0	−2 479	188.72	− 56 268
67	−2 427	213.92	− 64 108	300	63	−2 432	188.93	− 56 616
4 008	683	225.22	− 86 082	400	3 452	126	198.67	− 76 017
12 916	7 927	243.20	−133 000	600	10 498	5 509	212.93	−117 260
22 815	16 164	257.41	−183 110	800	17 991	11 340	223.69	−160 960
33 405	25 091	269.22	−235 810	1000	25 978	17 664	232.60	−206 620
44 484	34 507	279.31	−290 680	1200	34 476	24 499	240.33	−253 920
55 907	44 266	288.11	−347 440	1400	43 447	31 806	247.24	−302 690
67 580	54 277	295.90	−405 860	1600	52 844	39 541	253.51	−352 780
79 442	64 476	302.88	−465 750	1800	62 609	47 643	259.26	−404 060
91 450	74 821	309.21	−526 970	2000	72 689	56 060	264.57	−456 450
103 570	85 283	314.99	−589 400	2200	83 036	64 744	269.50	−509 860
115 790	95 833	320.30	−652 940	2400	93 604	73 650	274.10	−564 230
128 080	106 470	325.22	−717 490	2600	104 370	82 752	278.41	−619 490
140 440	117 160	329.80	−782 990	2800	115 290	92 014	282.45	−675 580
152 860	127 920	334.08	−849 390	3000	126 360	101 420	286.27	−732 460
165 330	138 720	338.11	−916 620	3200	137 550	110 950	289.88	−790 080
177 850	149 580	341.90	−984 620	3400	148 850	120 590	293.31	−848 390
190 410	160 470	345.49	−1 053 360	3600	160 250	130 320	296.57	−907 390
203 000	171 400	348.90	−1 122 800	3800	171 720	140 130	299.67	−967 010
215 630	182 370	352.13	−1 192 900	4000	183 280	150 020	302.63	−1 027 250

18

$\tilde{h}$ [kJ/kmol]	$\tilde{u}$ [kJ/kmol]	$\tilde{s}^{\ominus}$ [kJ/kmol K]	$\tilde{g}^{\ominus}$ [kJ/kmol]	T [K]	$\tilde{h}$ [kJ/kmol]	$\tilde{u}$ [kJ/kmol]	$\tilde{s}^{\ominus}$ [kJ/kmol K]	$\tilde{g}^{\ominus}$ [kJ/kmol]
Hydrogen (H_2)		$\tilde{m} = 2.016 \dfrac{kg}{kmol}$			Carbon Monoxide (CO)		$\tilde{m} = 28.0105 \dfrac{kg}{kmol}$	
−8 468	−8 468	0	− 8 468	0	−8 699	−8 669	0	− 8 669
−5 293	−6 124	102.04	− 15 496	100	−5 770	−6 601	165.74	− 22 344
−2 770	−4 433	119.33	− 26 635	200	−2 858	−4 521	185.92	− 40 041
0	−2 479	130.57	− 38 931	298.15	0	−2 479	197.54	− 58 898
54	−2 440	130.75	− 39 172	300	54	−2 440	197.72	− 59 263
2 958	− 368	139.11	− 52 684	400	2 975	− 351	206.12	− 79 475
8 812	3 823	150.97	− 81 769	600	8 941	3 953	218.20	− 121 980
14 703	8 051	159.44	− 112 850	800	15 175	8 524	227.16	− 166 550
20 686	12 371	166.11	− 145 430	1000	21 686	13 371	234.42	− 212 740
26 794	16 817	171.68	− 179 220	1200	28 426	18 449	240.56	− 260 250
33 062	21 422	176.51	− 214 050	1400	35 338	23 698	245.89	− 308 910
39 522	26 219	180.82	− 249 790	1600	42 384	29 081	250.59	− 358 560
46 150	31 184	184.72	− 286 350	1800	49 522	34 556	254.80	− 409 110
52 932	36 303	188.30	− 323 660	2000	56 739	40 110	258.60	− 460 460
59 860	41 569	191.60	− 361 650	2200	64 019	45 728	262.06	− 512 520
66 915	46 960	194.67	− 400 290	2400	71 346	51 391	265.25	− 565 260
74 090	52 473	197.54	− 439 510	2600	78 714	57 096	268.20	− 618 610
81 370	58 090	200.23	− 479 280	2800	86 115	62 835	270.94	− 672 530
88 743	63 799	202.78	− 519 590	3000	93 542	68 598	273.51	− 726 980
96 199	69 592	205.18	− 560 390	3200	101 000	74 391	275.91	− 781 930
103 740	75 469	207.47	− 601 650	3400	108 480	80 210	278.18	− 837 340
111 360	81 430	209.65	− 643 370	3600	115 980	86 044	280.32	− 893 190
119 060	87 469	211.73	− 685 510	3800	123 490	91 900	282.36	− 949 460
126 850	93 589	213.73	− 728 060	4000	131 030	97 769	284.29	−1 006 120
Oxygen (O_2)		$\tilde{m} = 31.999 \dfrac{kg}{kmol}$			Nitrogen (N_2)		$\tilde{m} = 28.013 \dfrac{kg}{kmol}$	
−8 682	−8 682	0	− 8 682	0	−8 669	−8 669	0	− 8 669
−5 778	−6 610	173.20	− 23 098	100	−5 770	−6 601	159.70	− 21 740
−2 866	−4 529	193.38	− 41 541	200	−2 858	−4 521	179.88	− 38 833
0	−2 479	205.03	− 61 131	298.15	0	−2 479	191.50	− 57 096
54	−2 440	205.21	− 61 509	300	54	−2 440	191.68	− 57 450
3 029	− 297	213.76	− 82 477	400	2 971	− 355	200.07	− 77 058
9 247	4 258	226.35	− 126 560	600	8 891	3 902	212.07	− 118 350
15 841	9 189	235.81	− 172 810	800	15 046	8 394	220.91	− 161 680
22 707	14 392	243.48	− 220 770	1000	21 460	13 145	228.06	− 206 600
29 765	19 788	249.91	− 270 120	1200	28 108	18 131	234.12	− 252 830
36 966	25 325	255.45	− 320 670	1400	34 936	23 296	239.38	− 300 190
44 279	30 976	260.34	− 372 260	1600	41 903	28 600	244.03	− 348 540
51 689	36 723	264.70	− 424 770	1800	48 982	34 016	248.19	− 397 770
59 199	42 571	268.65	− 478 110	2000	56 141	39 512	251.97	− 447 800
66 802	48 510	272.28	− 532 210	2200	63 371	45 079	255.41	− 498 540
74 492	54 537	275.63	− 587 010	2400	70 651	50 696	258.58	− 549 940
82 274	60 657	278.74	− 642 440	2600	77 981	56 364	261.51	− 601 950
90 144	66 864	281.65	− 698 490	2800	85 345	62 065	264.24	− 654 530
98 098	73 155	284.40	− 755 100	3000	92 738	67 795	266.79	− 707 640
106 130	79 521	286.99	− 812 240	3200	100 160	73 555	269.19	− 761 230
114 230	85 963	289.44	− 869 880	3400	107 610	79 339	271.45	− 815 310
122 400	92 467	291.78	− 928 010	3600	115 080	85 149	273.58	− 869 800
130 630	99 034	294.01	− 986 590	3800	122 570	90 976	275.60	− 924 730
138 910	105 660	296.13	−1 045 590	4000	130 080	96 819	277.53	− 980 040

Molar Properties of Some Gases and Vapours

$\tilde{h}$ [kJ/kmol]	$\tilde{u}$ [kJ/kmol]	$s°$ [kJ/kmol K]	$\tilde{g}°$ [kJ/kmol]	T [K]	$\tilde{h}$ [kJ/kmol]	$\tilde{u}$ [kJ/kmol]	$s°$ [kJ/kmol K]	$\tilde{g}°$ [kJ/kmol]
Hydroxyl (OH)		$\tilde{m} = 17.0075\ \frac{kg}{kmol}$			Nitric Oxide (NO)		$\tilde{m} = 30.006\ \frac{kg}{kmol}$	
− 9171	− 9171	0	− 9171	0	− 9192	− 9192	0	− 9192
− 6138	− 6969	149.48	− 21086	100	− 6071	− 6902	176.92	− 23763
− 2975	− 4638	171.48	− 37271	200	− 2950	− 4613	198.64	− 42678
0	− 2479	183.60	− 54740	298.15	0	− 2479	210.65	− 62806
54	− 2440	183.78	− 55080	300	54	− 2440	210.84	− 63198
3033	− 292	192.36	− 73909	400	3042	− 284	219.43	− 84729
8941	3953	204.33	− 113660	600	9146	4158	231.78	− 129920
14878	8227	212.87	− 115420	800	15548	8896	240.98	− 177240
20933	12618	219.62	− 198690	1000	22230	13915	248.43	− 226200
27158	17181	225.30	− 243200	1200	29121	19143	254.71	− 276540
33568	21928	230.23	− 288760	1400	36166	24526	260.14	− 328030
40150	26847	234.63	− 335250	1600	43321	30018	264.92	− 380550
46890	31924	238.59	− 382580	1800	50559	35594	269.18	− 433960
53760	37131	242.22	− 430670	2000	57861	41232	273.03	− 488190
60752	42460	245.55	− 479450	2200	65216	46924	276.53	− 543150
67839	47885	248.63	− 528870	2400	72609	52655	279.75	− 598780
75015	53397	251.50	− 578890	2600	80036	58418	282.72	− 655030
82266	58985	254.19	− 629460	2800	87492	64211	285.48	− 711860
89584	64640	256.71	− 680540	3000	94977	70034	288.06	− 769220
96960	70354	254.09	− 732130	3200	102480	75873	290.48	− 827070
104390	76118	261.34	− 784170	3400	110000	81733	292.77	− 885410
111860	81927	263.48	− 836670	3600	117550	87613	294.92	− 944170
119380	87783	265.51	− 889550	3800	125100	93507	296.96	− 1003360
126940	93680	267.45	− 942860	4000	132670	99417	298.90	− 1062950
Methane Vapour (CH$_4$)		$\tilde{m} = 16.043\ \frac{kg}{kmol}$			Ethylene Vapour (C$_2$H$_4$)		$\tilde{m} = 28.054\ \frac{kg}{kmol}$	
−10025	− 10025	0	− 10025	0	−10519	− 10519	0	− 10519
− 6699	− 7530	149.39	− 21638	100	− 7192	− 8024	180.44	− 25236
− 3368	− 5031	172.47	− 37863	200	− 3803	− 5466	203.85	− 44573
0	− 2479	186.15	− 55499	298.15	0	− 2479	219.22	− 65362
67	− 2427	186.37	− 55843	300	79	− 2415	219.49	− 65767
3862	536	197.25	− 75038	400	4883	− 1557	233.24	− 88412
13129	8141	215.88	− 116400	600	17334	12346	258.24	− 137610
24673	18022	232.41	− 161260	800	32849	26197	280.47	− 191520
38179	29865	247.45	− 209270	1000	50664	42350	300.30	− 249640
53271	43293	261.18	− 260150	1200	70254	60276	318.13	− 311510
69609	57969	273.76	− 313660	1400	91199	79558	334.27	− 376780
86910	73607	285.31	− 369590	1600	113180	99878	348.94	− 445120
104960	89994	295.93	− 427720	1800	135970	121010	362.36	− 516270
123600	106970	305.75	− 487900	2000	159390	142760	374.69	− 589990

The molar enthalpies of reaction, $\Delta\tilde{h}°$, on p. 21 are for a reference temperature of $T = 298.15\ K$ and are virtually independent of pressure. Corresponding values of Gibbs function of reaction, $\Delta\tilde{g}°$, may be found from values of equilibrium constant $K°$ using the relation

$$\Delta\tilde{g}°\,(= \tilde{g}°_P - \tilde{g}°_R) = -\tilde{R}T\ln K° \qquad \text{(Suffixes P and R refer to products and reactants)}$$

The *standard* or *thermodynamic equilibrium constant* is defined by

$$K° = \prod_i (p_i/p°)^{v_i} \quad \text{or} \quad \ln K° = \sum_i \ln(p_i/p°)^{v_i}$$

where v_i are the stoichiometric coefficients, those for the products and reactants being taken as positive and negative respectively. The constant so defined is dimensionless.

Reaction (kmol)— the values of $\Delta\tilde{h}^\circ$ relate to the corresponding chemical equation with amounts of substance in kilomoles	$\Delta\tilde{h}^\circ/[\text{kJ/kmol}]$ at $T = 298.15$ K (25 °C)
$C(sol) + O_2 \rightarrow CO_2$	− 393 520
$CO + \frac{1}{2}O_2 \rightarrow CO_2$	− 282 990
$H_2 + \frac{1}{2}O_2 \rightarrow H_2O(vap)$	− 241 830
$CH_4(vap) + 2O_2 \rightarrow CO_2 + 2H_2O(vap)$	− 802 310
$C_2H_4(vap) + 3O_2 \rightarrow 2CO_2 + 2H_2O(vap)$	− 1 323 170
$C_2H_6(vap) + 3\frac{1}{2}O_2 \rightarrow 2CO_2 + 3H_2O(vap)$	− 1 427 860
$C_6H_6(vap) + 7\frac{1}{2}O_2 \rightarrow 6CO_2 + 3H_2O(vap)$	− 3 169 540
$C_8H_{18}(vap) + 12\frac{1}{2}O_2 \rightarrow 8CO_2 + 9H_2O(vap)$	− 5 116 180
$CO_2 + H_2 \rightarrow CO + H_2O(vap)$	+ 41 160
$\frac{1}{2}H_2 + OH \rightarrow H_2O(vap)$	− 281 540
$\frac{1}{2}N_2 + \frac{1}{2}O_2 \rightarrow NO$	+ 90 290
$2H \rightarrow H_2$	− 435 980
$2O \rightarrow O_2$	− 498 340
$2N \rightarrow N_2$	− 945 300

At 298.15 K for H_2O $\tilde{h}_{fg} = 43\,990$ kJ/kmol of H_2O
for C_6H_6 $\tilde{h}_{fg} = 33\,800$ kJ/kmol of C_6H_6
for C_8H_{18} $\tilde{h}_{fg} = 41\,500$ kJ/kmol of C_8H_{18}

T [K]	\multicolumn{8}{c}{$\ln K^\circ$}
	$\dfrac{(p_{H_2O})(p^\circ)^{\frac{1}{2}}}{(p_{H_2})(p_{O_2})^{\frac{1}{2}}}$	$\dfrac{(p_{CO_2})(p^\circ)^{\frac{1}{2}}}{(p_{CO})(p_{O_2})^{\frac{1}{2}}}$	$\dfrac{(p_{H_2O})(p_{CO})}{(p_{H_2})(p_{CO_2})}$	$\dfrac{(p_{H_2O})(p^\circ)^{\frac{1}{2}}}{(p_{OH})(p_{H_2})^{\frac{1}{2}}}$	$\dfrac{(p_{NO})}{(p_{O_2})^{\frac{1}{2}}(p_{N_2})^{\frac{1}{2}}}$	$\dfrac{(p_{H_2})(p^\circ)}{(p_H)^2}$	$\dfrac{(p_{O_2})(p^\circ)}{(p_O)^2}$	$\dfrac{(p_{N_2})(p^\circ)}{(p_N)^2}$
298.15	92.207	103.762	− 11.554	106.329	− 34.933	164.005	186.961	367.479
300	91.604	103.057	− 11.453	105.627	− 34.707	162.922	185.723	365.126
400	67.321	74.669	− 7.348	77.360	− 25.655	119.164	135.710	270.329
600	42.897	46.245	− 3.348	48.956	− 16.602	75.226	85.519	175.356
800	30.592	32.036	− 1.444	34.670	− 12.072	53.135	60.319	127.753
1000	23.162	23.528	− 0.366	26.063	− 9.353	39.808	45.145	99.128
1200	18.182	17.871	0.311	20.307	− 7.541	30.878	35.005	80.011
1400	14.608	13.841	0.767	16.181	− 6.245	24.468	27.742	66.329
1600	11.921	10.829	1.091	13.086	− 5.273	19.637	22.285	56.055
1800	9.825	8.497	1.329	10.673	− 4.518	15.865	18.030	48.051
2000	8.145	6.634	1.510	8.741	− 3.912	12.840	14.622	41.645
2200	6.768	5.119	1.649	7.161	− 3.417	10.358	11.827	36.391
2400	5.619	3.859	1.759	5.844	− 3.005	8.281	9.497	32.011
2600	4.647	2.800	1.847	4.730	− 2.657	6.517	7.521	28.304
2800	3.811	1.893	1.918	3.774	− 2.360	5.002	5.826	25.117
3000	3.086	1.110	1.976	2.945	− 2.102	3.689	4.357	22.359
3200	2.450	0.429	2.022	2.220	− 1.877	2.538	3.072	19.936
3400	1.891	− 0.170	2.061	1.582	− 1.679	1.516	1.935	17.800
3600	1.391	− 0.702	2.093	1.016	− 1.504	0.609	0.926	15.898
3800	0.944	− 1.176	2.121	0.507	− 1.347	− 0.202	0.019	14.198
4000	0.541	− 1.600	2.141	0.051	− 1.207	− 0.934	− 0.796	12.660
4500	− 0.313	− 2.491	2.178	− 0.914	− 0.914	− 2.482	− 2.514	9.414
5000	− 0.997	− 3.198	2.201	− 1.683	− 0.682	− 3.725	− 3.895	6.807
5500	− 1.561	− 3.771	2.210	− 2.314	− 0.493	− 4.743	− 5.024	4.666
6000	− 2.033	− 4.246	2.213	− 2.839	− 0.338	− 5.590	− 5.963	2.865

$$p^\circ = 1 \text{ bar} = \frac{1}{1.01325} \text{ atm} \qquad (p^\circ)^{\frac{1}{2}} = 0.99344 \text{ atm}^{\frac{1}{2}} \qquad \log_{10} K^\circ = 0.43429 \ln K^\circ$$

A Selection of Chemical Thermodynamic Data

	$\tilde{m}$ [kg/kmol]	at $p^{\ominus} = 1$ bar and $T_0 = 298.15$ K				
		$\Delta \tilde{h}^{\ominus}_{f0}$ [kJ/kmol]	$\Delta \tilde{g}^{\ominus}_{f0}$ [kJ/kmol]	$\ln K^{\ominus}_{f0}$	$\tilde{c}^{\ominus}_{p0}$ [kJ/kmol K]	$\tilde{s}^{\ominus}_0$ [kJ/kmol K]
C (graphite)	12.011	0	0	0	8.53	5.69
C (diamond)	12.011	1 900	2 870	−1.157	6.06	2.44
C (gas)	12.011	714 990	669 570	−270.098	20.84	158.10
CH_4 (gas)	16.043	−74 870	−50 810	20.498	35.64	186.26
C_2H_4 (gas)	28.054	52 470	68 350	−27.573	42.89	219.33
CO (gas)	28.0105	−110 530	−137 160	55.331	29.14	197.65
CO_2 (gas)	44.010	−393 520	−394 390	159.093	37.13	213.80
H (gas)	1.008	217 990	203 290	−82.003	20.79	114.71
H_2 (gas)	2.016	0	0	0	28.84	130.68
OH (gas)	17.005	39 710	35 010	−14.122	29.99	183.61
H_2O (liq)	18.0155	−285 820	−237 150	95.660	75.32	70.00
H_2O (vap)	18.0155	−241 830	−228 590	92.207	33.58	188.83
N (gas)	14.0065	472 650	455 500	−183.740	20.79	153.30
N_2 (gas)	28.013	0	0	0	29.21	191.61
NO (gas)	30.006	90 290	86 600	−34.933	29.84	210.76
O (gas)	15.9995	249 170	231 750	−93.481	21.91	161.06
O_2 (gas)	31.999	0	0	0	29.37	205.14

Reproduced from Rogers, G. F. C., and Mayhew, Y. R., *Engineering Thermydynamics, Work and Heat Transfer* (Longman, 1992).

	$T/[K]$	250	300	400	500	600	800	1000
Ammonia (NH_3)	c_p	4.52	4.75	6.91	—	—	—	—
sat. liquid	ρ	669	600	346	—	—	—	—
t.p. = 195.4 K	$\mu/10^{-6}$	245	141	38	—	—	—	—
$\tilde{m}$ = 17.030 kg/kmol	$k/10^{-6}$	592	477	207	—	—	—	—
R–12 (CF_2Cl_2)	c_p	0.902	0.980	—	—	—	—	—
sat. liquid	ρ	1468	1304	—	—	—	—	—
t.p. = 115.3 K	$\mu/10^{-6}$	336	213	—	—	—	—	—
$\tilde{m}$ = 120.91 kg/kmol	$k/10^{-6}$	86.8	68.6	—	—	—	—	—
R–134a ($CH_2F–CF_3$)	c_p	1.297	1.426	—	—	—	—	—
sat. liquid	ρ	1367	1200	—	—	—	—	—
t.p. = 169.85 K	$\mu/10^{-6}$	369	194	—	—	—	—	—
$\tilde{m}$ = 102.03 kg/kmol	$k/10^{-6}$	104	83.4	—	—	—	—	—
Mercury (Hg)	c_p	0.141	0.139	0.137	0.137	0.137	0.138	—
liquid	ρ	13650	13530	13290	13050	12840	12420	—
m.p. = 234.3 K	$\mu/10^{-6}$	1880	1520	1190	1010	890	780	—
$\tilde{m}$ = 200.59 kg/kmol	k	0.0075	0.0081	0.0094	0.0107	0.0128	0.0137	—
Potassium (K)	$c_p/[kJ\ kg\ K]$		0.710	0.805	0.786	0.772	0.768	0.775
liquid	$\rho/[kg/m^3]$		860	812	789	766	721	675
m.p. 336.8 K	$\mu/10^{-6}[kg/m\ s]$	solid	417	319	258	179	133	
$\tilde{m}$ = 39.098 kg/kmol	$k/[kW/m\ K]$		0.099	0.0465	0.0454	0.0425	0.0337	0.0278
Sodium (Na)	c_p	1.179	1.224	1.369	1.315	1.277	1.273	1.277
liquid	ρ	977	967	921	897	872	823	774
m.p. 370.5 K	$\mu/10^{-6}$	solid	solid	610	420	320	230	180
$\tilde{m}$ = 22.990 kg/kmol	k	0.135	0.135	0.086	0.080	0.074	0.063	0.059
Sodium-Potassium	c_p	—	0.977	0.929	0.904	0.886	0.871	0.882
22%–78%	ρ	—	869	845	821	797	749	700
eutectic liquid	$\mu/10^{-6}$	solid	780	467	348	277	193	146
m.p. 262 K	k	—	0.0222	0.0236	0.0249	0.0262	0.0287	0.0312
Argon (Ar)	c_p	0.5203	0.5203	0.5203	0.5203	0.5203	0.5203	0.5203
1 atm	ρ	1.947	1.623	1.217	0.974	0.811	0.609	0.487
$\tilde{m}$ = 39.948 kg/kmol	$\mu/10^{-6}$	19.74	22.94	28.67	33.75	38.38	46.71	54.21
	$k/10^{-6}$	15.15	17.66	22.27	26.41	30.16	36.83	42.66
Carbon dioxide (CO_2)	c_p	0.791	0.846	0.939	1.014	1.075	1.169	1.234
1 atm	ρ	2.145	1.788	1.341	1.073	0.894	0.670	0.536
$\tilde{m}$ = 44.010 kg/kmol	$\mu/10^{-6}$	12.60	14.99	19.46	23.67	27.32	33.81	39.51
	$k/10^{-6}$	12.90	16.61	24.75	32.74	40.40	54.64	67.52
Helium (He)	c_p	5.193	5.193	5.193	5.193	5.193	5.193	5.193
1 atm	ρ	0.1951	0.1626	0.1220	0.0976	0.0813	0.0610	0.0488
$\tilde{m}$ = 4.003 kg/kmol	$\mu/10^{-6}$	18.40	20.80	25.23	29.30	33.12	40.19	46.70
	$k/10^{-6}$	134.0	149.8	177.9	202.6	224.7	—	—
Hydrogen (H_2)	c_p	14.05	14.31	14.48	14.51	14.55	14.69	14.98
1 atm	ρ	0.0983	0.0819	0.0614	0.0491	0.0409	0.0307	0.0246
$\tilde{m}$ = 2.016 kg/kmol	$\mu/10^{-6}$	7.92	8.96	10.87	12.64	14.29	17.34	20.13
	$k/10^{-6}$	156.1	181.7	228.1	271.8	314.7	402.2	—
Steam (H_2O)	c_p	1.855	1.864	1.901	1.954	2.015	2.147	2.288
low pressures	$\mu/10^{-6}$	—	9.42	13.2	17.3	21.3	29.5	37.6
$\tilde{m}$ = 18.015 kg/kmol	$k/10^{-6}$	—	18.8	26.6	35.7	46.3	70.8	97.9

* See footnote on p. 6.

The properties c_p, μ and k (and ρ for liquids) do not vary much with pressure; see also footnote on p. 16.

International Standard Atmosphere

z [m]	p [bar]	T [K]	ρ/ρ_0	v $10^{-5}[\mathrm{m^2/s}]$	k $10^{-5}[\mathrm{kW/m\,K}]$	a [m/s]	λ $10^{-8}[\mathrm{m}]$
− 2 500	1.3521	304.4	1.2631	1.207	2.661	349.8	5.251
− 2 000	1.2778	301.2	1.2067	1.253	2.636	347.9	5.497
− 1 500	1.2070	297.9	1.1522	1.301	2.611	346.0	5.757
− 1 000	1.1393	294.7	1.0996	1.352	2.585	344.1	6.032
− 500	1.0748	291.4	1.0489	1.405	2.560	342.2	6.324
0	1.01325	288.15	1.0000	1.461	2.534	340.3	6.633
500	0.9546	284.9	0.9529	1.520	2.509	338.4	6.961
1 000	0.8988	281.7	0.9075	1.581	2.483	336.4	7.309
1 500	0.8456	278.4	0.8638	1.646	2.457	334.5	7.679
2 000	0.7950	275.2	0.8217	1.715	2.431	332.5	8.072
2 500	0.7469	271.9	0.7812	1.787	2.405	330.6	8.491
3 000	0.7012	268.7	0.7423	1.863	2.379	328.6	8.936
3 500	0.6578	265.4	0.7048	1.943	2.353	326.6	9.411
4 000	0.6166	262.2	0.6689	2.028	2.327	324.6	9.917
4 500	0.5775	258.9	0.6343	2.117	2.301	322.6	10.46
5 000	0.5405	255.7	0.6012	2.211	2.275	320.5	11.03
5 500	0.5054	252.4	0.5694	2.311	2.248	318.5	11.65
6 000	0.4722	249.2	0.5389	2.416	2.222	316.5	12.31
6 500	0.4408	245.9	0.5096	2.528	2.195	314.4	13.02
7 000	0.4111	242.7	0.4817	2.646	2.169	312.3	13.77
7 500	0.3830	239.5	0.4549	2.771	2.142	310.2	14.58
8 000	0.3565	236.2	0.4292	2.904	2.115	308.1	15.45
8 500	0.3315	233.0	0.4047	3.046	2.088	306.0	16.39
9 000	0.3080	229.7	0.3813	3.196	2.061	303.8	17.40
9 500	0.2858	226.5	0.3589	3.355	2.034	301.7	18.48
10 000	0.2650	223.3	0.3376	3.525	2.007	299.5	19.65
10 500	0.2454	220.0	0.3172	3.706	1.980	297.4	20.91
11 000	0.2270	216.8	0.2978	3.899	1.953	295.2	22.27
11 500	0.2098	216.7	0.2755	4.213	1.952	295.1	24.08
12 000	0.1940	216.7	0.2546	4.557	1.952	295.1	26.05
12 500	0.1793	216.7	0.2354	4.930	1.952	295.1	28.18
13 000	0.1658	216.7	0.2176	5.333	1.952	295.1	30.48
13 500	0.1533	216.7	0.2012	5.768	1.952	295.1	32.97
14 000	0.1417	216.7	0.1860	6.239	1.952	295.1	35.66
14 500	0.1310	216.7	0.1720	6.749	1.952	295.1	38.57
15 000	0.1211	216.7	0.1590	7.300	1.952	295.1	41.72
15 500	0.1120	216.7	0.1470	7.895	1.952	295.1	45.13
16 000	0.1035	216.7	0.1359	8.540	1.952	295.1	48.81
16 500	0.09572	216.7	0.1256	9.237	1.952	295.1	52.79
17 000	0.08850	216.7	0.1162	9.990	1.952	295.1	57.10
17 500	0.08182	216.7	0.1074	10.805	1.952	295.1	61.76
18 000	0.07565	216.7	0.09930	11.686	1.952	295.1	66.79
18 500	0.06995	216.7	0.09182	12.639	1.952	295.1	72.24
19 000	0.06467	216.7	0.08489	13.670	1.952	295.1	78.13
19 500	0.05980	216.7	0.07850	14.784	1.952	295.1	84.50
20 000	0.05529	216.7	0.07258	15.989	1.952	295.1	91.39
22 000	0.04047	218.6	0.05266	22.201	1.968	296.4	126.0
24 000	0.02972	220.6	0.03832	30.743	1.985	297.7	173.1
26 000	0.02188	222.5	0.02797	42.439	2.001	299.1	237.2
28 000	0.01616	224.5	0.02047	58.405	2.018	300.4	324.0
30 000	0.01197	226.5	0.01503	80.134	2.034	301.7	441.3
32 000	0.00889	228.5	0.01107	109.62	2.051	303.0	599.4

Density at sea level $\rho_0 = 1.2250 \, \mathrm{kg/m^3}$.

SI – British Conversion Factors

The International System of Units (HMSO, 1986) may be consulted for the definitions of SI units, and *British Standard* 350 for comprehensive tables of conversion factors.

Exact values are printed in **bold type**.

Mass: $1 \text{ kg} = \dfrac{1}{\textbf{0.453 592 37}} \text{ lb} = 2.205 \text{ lb}$

Length: $1 \text{ m} = \dfrac{1}{\textbf{0.3048}} \text{ ft} = 3.281 \text{ ft}$

Volume: $1 \text{ m}^3 = 10^3 \text{ dm}^3 \text{ (litre)} = 35.31 \text{ ft}^3 = 220.0 \text{ UK gal} = 264.2 \text{ US gal}$

Time: $1 \text{ s} = \dfrac{1}{60} \text{ min} = \dfrac{1}{3600} \text{ h}$

Temperature unit: $1 \text{ K} = \textbf{1.8} \text{ R}$ (see p. 11 for definitions of units and scales)

Force: $1 \text{ N (or kg m/s}^2) = \mathbf{10^5} \text{ dyn} = \dfrac{1}{\textbf{9.806 65}} \text{ kgf}$

$$= 7.233 \text{ pdl} = \dfrac{7.233}{32.174} \text{ or } 0.2248 \text{ lbf}$$

Pressure: p: $1 \text{ bar} = \mathbf{10^5} \text{ N/m}^2 \text{ (or Pa)} = 14.50 \text{ lbf/in}^2 = 750 \text{ mmHg} = 10.20 \text{ mH}_2\text{O}$

Specific volume v: $1 \text{ m}^3/\text{kg} = 16.02 \text{ ft}^3/\text{lb}$

Density ρ: $1 \text{ kg/m}^3 = 0.062 43 \text{ lb/ft}^3$

Energy: $1 \text{ kJ} = 10^3 \text{ N m} = \dfrac{1}{\textbf{4.1868}} \text{ kcal}_{\text{IT}} = 0.9478 \text{ Btu} = 737.6 \text{ ft lbf}$

Power: $1 \text{ kW} = \mathbf{1 \text{ kJ/s}} = \dfrac{10^3}{\textbf{9.806 65}} \text{ kgf m/s} = \dfrac{10^3}{\textbf{9.806 65} \times \textbf{75}} \text{ metric hp}$

$$= 737.6 \text{ ft lbf/s} = \dfrac{737.6}{\textbf{550}} \text{ or } \dfrac{1}{0.7457} \text{ British hp} = 3412 \text{ Btu/h}$$

Specific energy etc. (u, h): $1 \text{ kJ/kg} = \dfrac{1}{\textbf{2.326}} \text{ Btu/lb} = 0.4299 \text{ Btu/lb}$

Specific heat capacity etc. (c, R, s): $1 \text{ kJ/kg K} = \dfrac{1}{\textbf{4.1868}} \text{ Btu/lb R} = 0.2388 \text{ Btu/lb R}$

Thermal conductivity k: $1 \text{ kW/m K} = 577.8 \text{ Btu/ft h R}$

Heat transfer coefficient: $1 \text{ kW/m}^2 \text{ K} = 176.1 \text{ Btu/ft}^2 \text{ h R}$

Dynamic viscosity μ: $1 \text{ kg/m s} = \mathbf{1 \text{ N s/m}^2} = 1 \text{ Pa s} = \mathbf{10} \text{ dyn s/cm}^2 \text{ (or poise)}$
$$= 2419 \text{ lb/ft h} = 18.67 \times 10^{-5} \text{ pdl h/ft}^2$$

Kinematic viscosity v: $1 \text{ m}^2/\text{s} = \mathbf{10^4} \text{ cm}^2/\text{s} \text{ (or stokes)} = 38\,750 \text{ ft}^2/\text{h}$

General Information

Standard acceleration: $g_n = \textbf{9.806\,65}\,\text{m/s}^2 = 32.1740\,\text{ft/s}^2$

Standard atmospheric pressure: $1\,\text{atm} = \textbf{1.013\,25}\,\text{bar}$

$$= 760\,\text{mmHg} = 10.33\,\text{mH}_2\text{O} = 1.0332\,\text{kgf/cm}^2$$
$$= 29.92\,\text{inHg} = 33.90\,\text{ftH}_2\text{O} = 14.696\,\text{lbf/in}^2$$

Molar (universal) gas constant: $\tilde{R} = 8.3145\,\text{kJ/kmol K}$†
$$= 1.986\,\text{Btu/lb-mol R} = 1545\,\text{ft lbf/lb-mol R}$$

Molar volume: $\tilde{v} = 22.41\,\text{m}^3/\text{kmol at 1 atm and 0°C}$
$$= 359.0\,\text{ft}^3/\text{lb-mol at 1 atm and 32°F}$$

Composition of air:

	vol. analysis	grav. analysis
Nitrogen ($N_2 - 28.013\,\text{kg/kmol}$)	0.7809	0.7553
Oxygen ($O_2 - 31.999\,\text{kg/kmol}$)	0.2095	0.2314
Argon ($Ar - 39.948\,\text{kg/kmol}$)	0.0093	0.0128
Carbon dioxide ($CO_2 - 44.010\,\text{kg/kmol}$)	0.0003	0.0005

Molar mass $\tilde{m} = 28.96\,\text{kg/kmol}$
Specific gas constant $R = 0.2871\,\text{kJ/kg K}$ See p. 16 for other properties
$= 0.068\,56\,\text{Btu/lb R} = 53.35\,\text{ft lbf/lb R}$

For approximate calculations with air:

	vol. analysis	grav. analysis
$N_2 - 28\,\text{kg/kmol}$	0.79	0.767
$O_2 - 32\,\text{kg/kmol}$	0.21	0.233
N_2/O_2	3.76	3.29

Molar mass $\tilde{m}$ $= 29\,\text{kg/kmol}$
Specific gas constant $R = 0.287\,\text{kJ/kg K}$
    $= 0.0685\,\text{Btu/lb R} = 53.3\,\text{ft lbf/lb R}$
$c_p = 1.005\,\text{kJ/kg K}\quad = 0.240\,\text{Btu/lb R}$
$c_v = 0.718\,\text{kJ/kg K}\quad = 0.1715\,\text{Btu/lb R}$
$c_p/c_v = \gamma = 1.40$

The Stefan-Boltzmann constant:
$$\sigma = 56.7 \times 10^{-12}\,\text{kW/m}^2\,\text{K}^4 = 0.171 \times 10^{-8}\,\text{Btu/ft}^2\,\text{h}\,\text{R}^4$$

† The kilomole (kmol) is the amount of substance of a system which contains as many elementary entities as there are atoms in 12 kg of carbon 12.
   The elementary entities must be specified, but for problems involving mixtures of gases and combustion they will be molecules or atoms.

PRINCIPAL SOURCES

National Engineering Laboratory, *Steam Tables 1964* (Her Majesty's Stationery Office 1964) (reproduced by courtesy of the Controller of Her Majesty's Stationery Office).

ASHRAE Thermodynamic Properties of Refrigerants and *ASHRAE Thermophysical Properties of Refrigerants* (ASHRAE, 1969 and 1976).

Hilsenrath, J., Beckett, C. W., Benedict, W. S., Fano, L., Hoge, H. J., Masi, J. F., Nuttall, R. L., Touloukian, Y. S., and Woolley, H. W., *Tables of Thermal Properties of Gases* (US. N.B.S. Circular 564, 1955, available from the Superintendent of Documents, Government Printing Office, Washington 25, D.C.).

Rossini, F. D., Wagman, D. D., Evans, W. H., Levine, S., and Jaffe. I., *Selected Values of Chemical Thermodynamic Properties* (ibid. 500, 1952).

Wagman, D. D., *Selected Valued of Chemical Thermodynamic Properties* (ibid., Tech. Note 270, 1965).

Sheldon, L. A., *Thermodynamic Properties of Mercury Vapour*, Amer. Soc. Mech. Engrs. No. 49–A–30 (1949).

Stull, D. R., and Prophet, H. (eds). *Janaf Thermochemical Tables*, The Dow Chemical Company, Midland, Michigan (U.S. G.P.O., 1971, Supplements 1974, 1975, 1978).

Thermodynamic Properties of KLEA 134a (ICI private communication, 1994)

By the same authors
Engineering Thermodynamics, Work and Heat Transfer (Longmans, 4th ed.).

By H. Cohen, G. F. C. Rogers and H. I. H. Saravanamuttoo
Gas Turbine Theory (Longmans, 4th ed.)

FOR USE WITH THESE TABLES

Enthalpy–Entropy Diagram for Steam
Prepared by D. C. Hickson *and* F. R. Taylor

NOTES